MAN IS becoming increasingly aware that he must pay attention to his environment if he is to avoid ruining the world completely. *Pacific Marine Life,* by Charles and Diana DeLuca, surveys the invertebrates that inhabit the Pacific Ocean and gives the layman an understanding of how important these creatures are to our environment. Carefully organized and clearly written, this very readable handbook was created especially for people who have no training in marine science, but who are interested in the sea.

It is a fascinating world the DeLucas explore. They examine the spotted octopus, whose salivary glands "contain enough venom to kill eight or ten men," and the sea cucumber, the vacuum cleaner of the ocean floor. The reader is also given a look at the Crown of Thorns, a starfish which received publicity several years ago as the destroyer of great stretches of reef. The range of animals is large. The reader is taken from the Chambered Nautilus to the giant Spider Crab to plankton.

The many black-and-white drawings included in the book help the reader to recognize the animals. The descriptions of the organisms' habits, habitats, edibility, and function in the environment make real what to many may have been only an exotic kingdom.

PACIFIC MARINE LIFE

PACIFIC MARINE LIFE

A Survey of Pacific Ocean Invertebrates

BY

CHARLES J. DeLUCA

Curator, Waikiki Aquarium

and

DIANA MACINTYRE DeLUCA

University of Hawaii

CHARLES E. TUTTLE COMPANY

Rutland, Vermont Tokyo, Japan

Representatives

For Continental Europe:
BOXERBOOKS, INC., *Zurich*

For the British Isles:
PRENTICE-HALL INTERNATIONAL, INC., *London*

For Canada:
HURTIG PUBLISHERS, *Edmonton*

For Australasia:
BOOK WISE (AUSTRALIA) PTY., LTD.
104-108 Sussex Street, Sydney

Published by the Charles E. Tuttle Company, Inc.
of Rutland, Vermont and Tokyo, Japan
with Editorial offices at
Suido 1-chome, 2-6 Bunkyo-ku, Tokyo

Library of Congress Catalog Card No. 76-12228
International Standard Book No. 0-8048 1212-8

First printing, 1976

Printed in Japan

Dedicated to
SPENCER WILKIE TINKER,
Pacific naturalist

CONTENTS

FOREWORD

This small volume on marine life concentrates on the invertebrates only and is intended as a survey for those people interested in learning about the diverse life forms to be found in the broad Pacific Ocean. The material included here represents a series of nature pamphlets published over many years by the Waikiki Aquarium of the University of Hawaii as educational guides for visitors of all ages; therefore, some of the scientific names of the animals may have changed as well as some of the information about them.

It is hoped, however, that what is included in these pages will instill in the casual reader an appreciation of the animals which too often are viewed simply as uninspiring and "low" forms of life, and a realization of their complexity as well as of their importance to the planet and to man.

Aleutian Is.
JAPAN
Midway
Hawaii
Wake I.
Mariana Is.
PHILIPPINES
Guam
Palmyra I.
MALAYSIA
New Guinea
Gilbert Is.
Galapagos Is.
INDONESIA
Samoa
Tonga
Tahiti
PACIFIC OCEAN
AUSTRALIA
NEW ZEALAND
Tasmania

INTRODUCTION

The Pacific Ocean is the largest in the world and while it is usually thought of as warm and tropical it must be kept in mind that the waters sweep along the coast of Asia until they join those of the Arctic Ocean, and along the coast of South America from the cold Antarctic. Like the Atlantic, then, the Pacific has diverse areas of water temperature, in the case of the latter ranging from freezing to a possible 81 degrees F. on the coast of Australia, each area supporting different life forms. Such great diversity is part of what gives this mighty ocean its unique and mysterious appeal to both scientists and laymen.

The Pacific, probably more than any other ocean, seems to offer unlimited scope for marine study. Named by Magellan when he navigated it in 1520, the Pacific lives up to its name by being tranquil for large parts of the year and so offers easy access to research vessels. Most active in the exploration of the ocean have been the Americans, Russians, and Japanese, who have sought to map the numerous ridges, fracture zones, fault lines, volcanoes, guyots (flat-topped seamounts lying below the surface), and trenches that dot the floor. This research as a whole has shown that the Pacific is incredibly deep. The lowest depth yet recorded is located near the Marianas Trench in the western Pacific where a depth of 36,000 feet was discovered; this is the deepest spot yet found in any ocean. By comparison, one of the highest points on earth is the 29,002 ft. height of Mt. Everest in the Himalayas. An even more impressive contrast between the land and sea lies in the fact that all the oceans of the world combined cover an area more than twice that of the land masses; the Pacific alone covers over 63,000,000 square miles and with its volume of 170,000,000 cubic miles it could comfortably contain all the surface land masses and still have room left over for another continent the size of Africa.

The physical study of this vast area has been much changed in recent years. There have been new developments in photography, sounding, heat flow, and magnectic detection. New theories have emerged, including the concepts of continental drift, sea-floor spread, and plate tectonics (the belief that the earth's crust is made up of plates that are created at one end and destroyed at another). In the necessary testing and research to support these ideas the Pacific figures importantly.

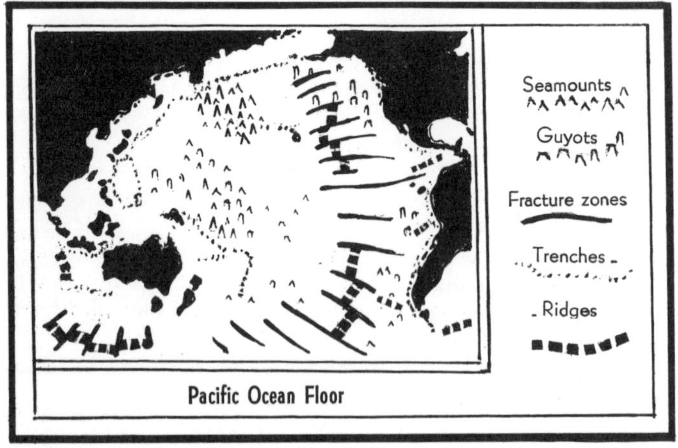

Pacific Ocean Floor

Marine geologists are not the only scientists interested in the Pacific. Marine zoologists find many unique forms of life a fruitful source of study. Nowhere else is to be found such creatures as the crown of thorns star-fish (Hawaii to the Red Sea), the chambered nautilus (Philippines, New Caledonia, Fiji), the conus gloria-maris shell (Solomons, Philippines, Fiji), the coconut crab (Palmyra to the Indian Ocean), the spotted or blue-ringed octopus (Australia), the giant clam (Australia, Marshalls), the giant spider crab (Japan), the king crab (Japan and Alaska), the giant octopus (Puget Sound), the pismo clam (Southern California).

Among the most fascinating of the animals that inhabit the Pacific are those which created the coral islands, reefs, and atolls. The tiny polyp responsible for these masterpieces is often measured in fractions of an inch; it varies with the species, but one-half inch per year is good average growth. It takes sixteen years then to build a coral head eight inches across. The construction of the Great Barrier Reef, the Pacific's and the world's largest reef, took thousands, perhaps millions of years, and its volume is many times over that of every existing building constructed by man.

The Pacific has generated controversies, scientific and quasi-scientific, as any mysterious area must. One is over a stretch of ocean near Japan which has claimed shipping in much the same way as the Bermuda triangle. In the face of the unknown, man can only guess at the causes. Another is over the formation of the ocean itself. Many people still believe that the moon was torn out of the earth leaving behind the hollow between the continents of Asia and North America. This has been generally discredited by scientists who point out that the moon is more than thirty times larger than the Pacific. But old ideas die hard and such fictions give a romantic aura to the ocean.

The ancient Hawaiians, like other Polynesian peoples, had many chants and legends about the sea and since the advent of Europeans the Pacific has figured in the literature of the West as well as that of the East. One remembers here the work of Herman Melville, Jack London, Robert Louis Stevenson, Joseph Conrad, Nordhoff and Hall, Yukio Mishima, and James Michener, to name but a few. It has cast its spell on artists too, numbering Gauguin among those who have been fascinated by the sea and the peoples living beside it. Obviously, an ocean as large as this has enough to satisfy both those who deal in facts and those who prefer to hear of stories and of mysteries.

Unlike Magellan, people no longer look for a passage to the East by circumnavigating the world in wooden ships; travel across the Pacific has been cut from months to a matter of hours.

But however traversed, this ocean will always have a special fascination for both the visitor and those who make their home on its shores, offering to the scientist and layman alike an unlimited laboratory of unique geological formations and unmatched zoological specimens.

LIFE IN THE SEA

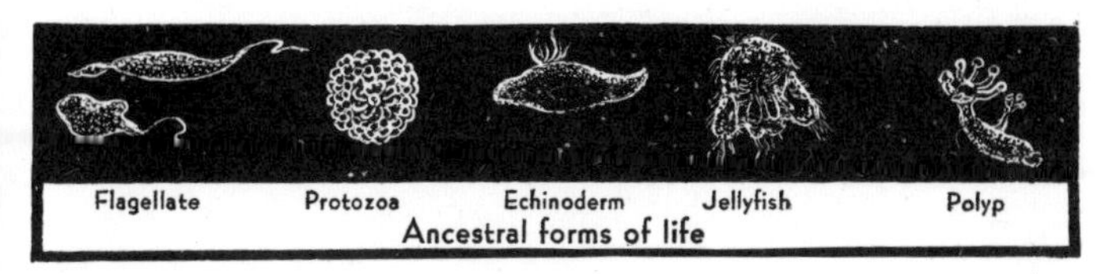

Ancestral forms of life

There is evidence supporting the belief that the first animals began to develop on earth several million years before the Cambrian Period (600 to 500 million years ago). Biologists are certain that animals during pre-Cambrian times possessed sensory organs, digestive systems and nervous systems even though there was, as yet, neither protective bone nor shell.

The creature today which is probably most like the common relative of all life is an important microscopic lifeform that is half plant-half animal. It is believed that this lifeform eventually produced the ***Protozoa,*** the phylum of one-celled animals. Protozoans, in turn, were able to develop the capacity for sexual reproduction, digestion and locomotion.

With the advent of multicellular organisms, animal forms more complex and varied became possible, creatures such as polyps, jellyfishes and the first of the sponges. Also during the Pre-Cambrian Period the ancestral echinoderm developed, and from the echinoderm two lines formed, one leading to the starfish, urchins and the like, the other resulting in the animals which inherited the earth, the vertebrates.

During the Cambrian Period, the most important development was that of protective shell, plate and skin. A low form of sponge enjoyed a brief but prolific life, but was replaced by the brachiopods (lampshells) and, eventually, the highly successful trilobite.

The Cambrian Period gave way to the Ordovician (500 to 425 million years ago). The trilobite lost its hegemony of the sea to the nautiloid — a mollusk that is related to today's octopus. The Giant Nautiloid of that age was protected by a shell 15 feet long.

It was during the Ordovician Period that coral developed, as did the clam and starfish.

The transition between the Ordovician Period and the Silurian (425 to 405 million years ago) saw no great or immediate change. The trilobite continued its decline. The clams, corals, sponges and snails continued to thrive and multiply. Nautiloids multiplied too, but the new giant of the sea became the Sea Scorpion, the size of which ranged from several inches to eight or nine feet.

The first true vertebrates were armored, jawless fish which swam from place to place sifting microscopic nutrients from the water and from the muddy bottom. The first jawed fish, the *Acanthodian*, developed in the late Silurian Period and, though only a few inches in length, was a predator, probably sharklike in appearance and definitely the ancestor of the fish we know today.

Acanthodian

At the same time marine plants began to adapt to the land masses. During the Devonian Period which followed, the low-lands became well carpeted with leafy plants and ferns. It was this hospitable environment which sustained the first exodus of animals from the sea, that of the arthropods.

The Lungfish developed its respiratory system, and the first amphibian, the *Ichthyostega*, was created, an animal with a fish-like tail and fin-like feet.

Life on land was well under way.

PROTOZOANS

Phylum *Protozoa*

The distinguishing feature of *Protozoa* is its single-celled body. It is not divided into organs, tissues and the like, and, so, is described as uni-cellular. Protozoa's place in the "tree of life" is probably best described by its name; *proto* -first, *zoon* -animal, from the Greek. Protozoans are microscopic in size.

The group is a large one which is found in marine, brackish or fresh water, and is varied in its functional makeup depending on the particular environment of its many families. Some are sedentary while others are capable of limited movement. It is the manner of locomotion that is one of the more important features of protozoan classification. Many are referred to as flagellates or ciliates, so named because of the hair-like flagella or cilia that can be seen projecting from the animal. These appendages beat swiftly and cause the Protozoa to move through the water. The difference between the two is difficult to determine in many cases. Generally, however, cilia are a great deal shorter than flagella, are usually arranged in rows, and are more numerous. Also, groups of cilia appear to have a uniform beat while flagella display a more independent, uncoordinated movement.

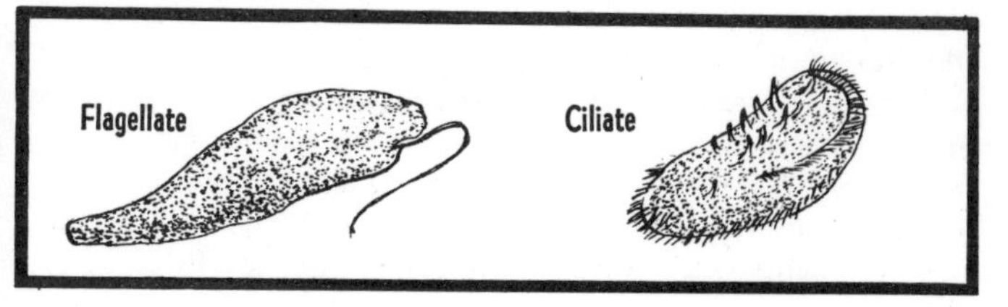

Another group of ***Protozoa*** is the Amoebae. They depend neither on cilia nor flagella for locomotion, but on a kind of "foot". They are classified pseudopodia ("false foot") along with several other ***Protozoa***. They move by forming temporary finger-like protrusions and progress in a slow, fluid manner.

Forms of Amoebae

Protozoa, depending on the type, get their nutrition in various ways—some by photosynthesis, others by absorption of dissolved material in the environment. Most ***Protozoa***, however, ingest solid food such as bacteria, waste or, even, other ***Protozoa***. Scientists look upon the group that depends on photosynthesis for nutrition as a kind of natural link between the plant and animal worlds.

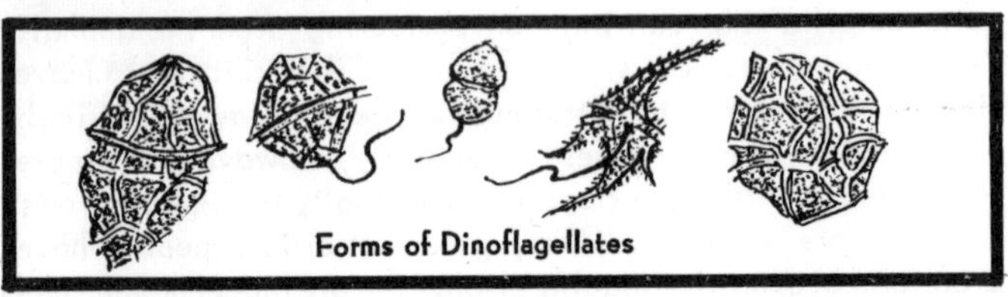

Forms of Dinoflagellates

Dinoflagellates are particularly interesting because some of them are responsible for "red tides", the end products of which are mass bodies of decayed matter that in turn poison the environment. And, it is also dinoflagellates that cause the often spectacular displays of luminescence seen by most ocean travellers. The animals sparkle briefly when disturbed and even people wandering along the wet sand in shoreline areas at night notice glowing spots of light surrounding each footfall.

PORIFERA

Phylum *Porifera*

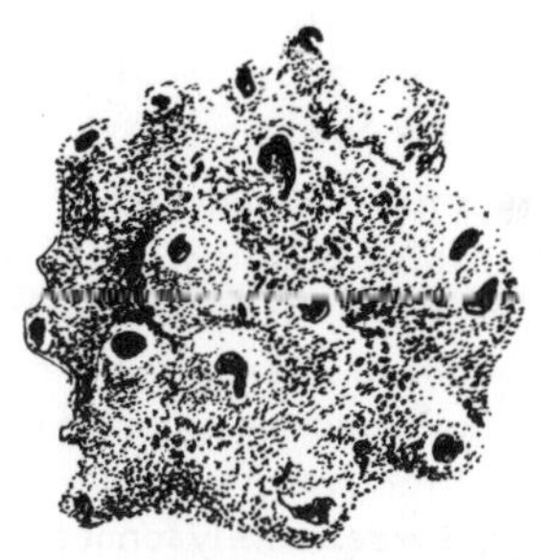

The lowest forms of the multi-celled animals are the Sponges (Phylum ***Porifera***). Their distribution ranges from fresh water to salt water, although most occur in the oceans. Many are shoreline and reef inhabitants but some develop in deep water. A sponge is very porous; its body walls are provided with canals through which water moves. A system of cells enables the animal to control this flow of water and the nutritious organisms suspended in it; the cells also allow the animal to provide itself with the necessary oxygen.

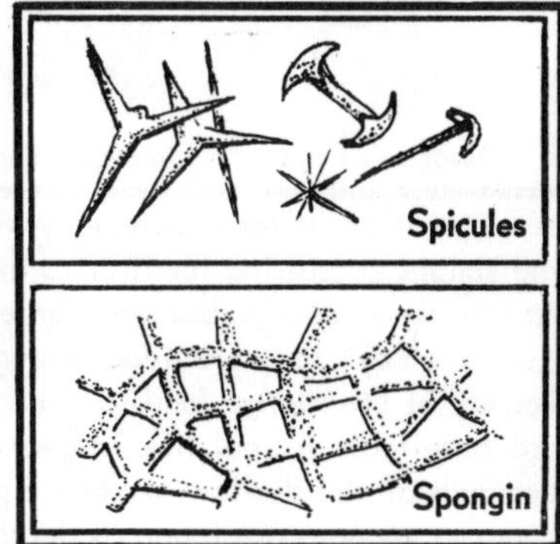

The supporting skeleton of the sponge may be comprised of spicules, spongin, or a combination of both. Spicules are calcareous or siliceous formations which, like our own skeletons, fashion a supporting framework for living tissues. Spongin is a fibrous material which remains elastic after the death of the animal. In commercial bath sponges, the skeletons are made up of soft but tough spongin. Today synthetic sponges have all but wiped out the demand for the authentic variety, but years ago real sponges were a multi-million dollar industry. This was especially true in the Caribbean Sea.

A sponge industry was never developed in the Pacific. This in no way suggests that this ocean is without an ample supply of sponges. On the contrary, there are many types ranging from the colorful Redbeard to the curiously named Deadman's Finger. Vase sponges have become popular among collectors as has one of the most attractive of all nautical creatures, the Venus Flowerbasket. It is a delicately formed sponge the skeleton of which is made of silica, imparting to it the appearance of woven glass when properly cleaned and mounted.

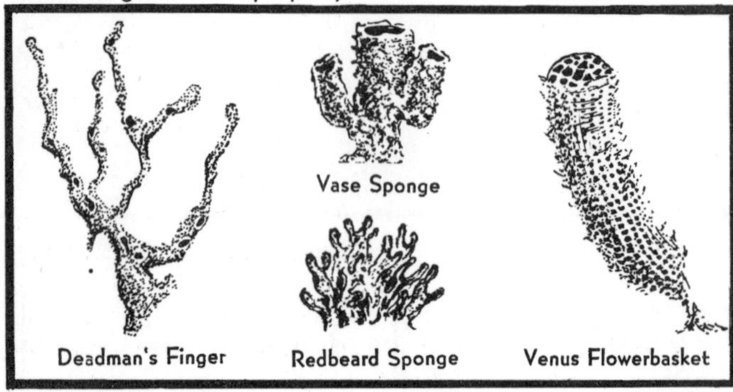

Hawaiian waters have many kinds of sponges ranging from the simple encrusting families to the large branching varieties. At one time Pearl Harbor was carpeted by colonies of Porifera. Kaneohe Bay, too, provided enough shelter to sustain a wide variety of the sponge family. The very same conditons which are hospitable to reef life in general appear to satisfy the demands of the shallow water sponges. The larger, more compact sponges are found at depths of at least 200 feet.

ANNELIDS

Phylum *Annelida*

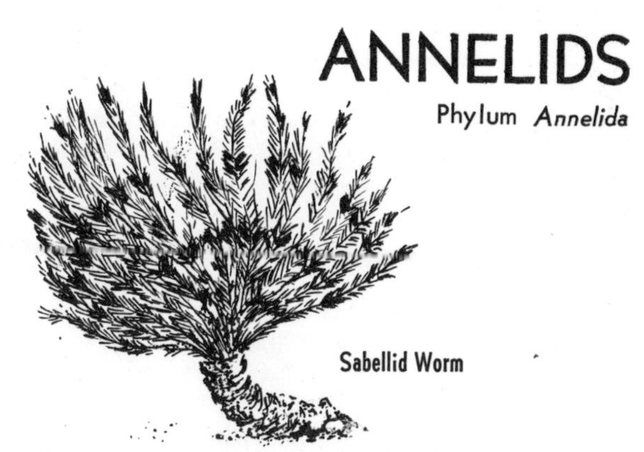

Sabellid Worm

Annelid worms are an important phylum of the animal kingdom. They are far better developed than many other invertebrates, and have specialized digestive tracts and excretory systems, advanced nervous systems and sense organs. One of their more important features is their segmentation both internally and externally.

The annelids include earthworms and leeches as well as the interesting class of marine animals known as Polychaetes. The latter group embraces creatures that may be either free-moving or burrowing. The free-moving polychaetes inhabit the open ocean while their burrowing cousins prefer to live on the bottom and so do not cover as much nautical territory in their lifetime. Sedentary polychaetes live in tubes which they construct themselves.

A number of annelid worms, including some pelagic (open ocean) polychaetes, have light-producing capabilities.

The polychaetes, comprising over sixty-five families, are the largest class of annelids. While a few live in brackish to fresh water, and some in moist soil, most inhabit the marine environment. A number of families appear to favor tropical areas while others are found in colder climates.

Perhaps the most attractive of the Polychaetes is the Sabellid worm, especially noteworthy because of its flower-like appearance. It is enclosed in a mucilagenous tube into which it quickly withdraws when alarmed. The sabellid has a pair of gill tufts on its head which branch into a number of filaments. These filaments resemble feathers, serve as respiratory organs and, in some species, contain eyes on their outer edges.

Terebellid worms inhabit tubes which are composed of the animal's mucus secretions together with sand, shells and small chips of stone. Some large specimens may reach 12 inches in length.

The serpulid worms form calcereous tubes as habitats. A member of this group is able to close out the world by means of a kind of hatch cover (operculum) which seals his tube after he withdraws into it. Specimens attain a length of about 3 inches and are usually found attached to stone, coral or driftwood.

The spaghetti worm, *Lanice conchilaga*, known to Hawaiians as kaunaoa, is usually found on reefs or in tide pools. Usually its body is hidden in a crevice with only its tentacles showing. An extract from the tentacles has been under study as a cancer fighter.

Terebellid worm | Serpulid worm | Spaghetti worm

COELENTERATES

Phylum *Coelenterata*

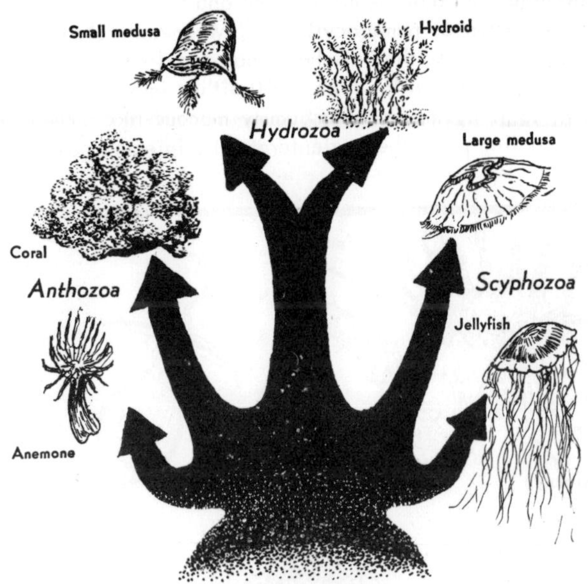

There are two theories dealing with the evolution of coelenterates. One holds that they have descended from colonial protozoans, while the second maintains that primitive flatworms gave rise to the phylum. While neither theory appears to have enough supporting evidence to be universally accepted, both provide enough challenges to keep scientists delving into the coelenterate's past.

It is generally accepted that the coelenterates were the first animals to develop definite tissues. They are two-layered creatures (ectoderm and endoderm), with a central digestive cavity and an elaborate system of stinger-bearing tentacles.

Coelenterates are also known as cnidarians (after the Greek word for thread, ***cnidos***), an apt name because of the stinging cells. These are located by the hundreds on each tentacle and serve as both defense mechanisms and food gatherers. The cells are known as nematocysts and, when activated by prey such as small shrimp, they discharge harpoon-like, spiny threads which, depending on the type, will either impale the victim or stick to it by means of a gummy mucous-like substance. When impaling prey, the coelenterate also injects a paralyzing toxin.

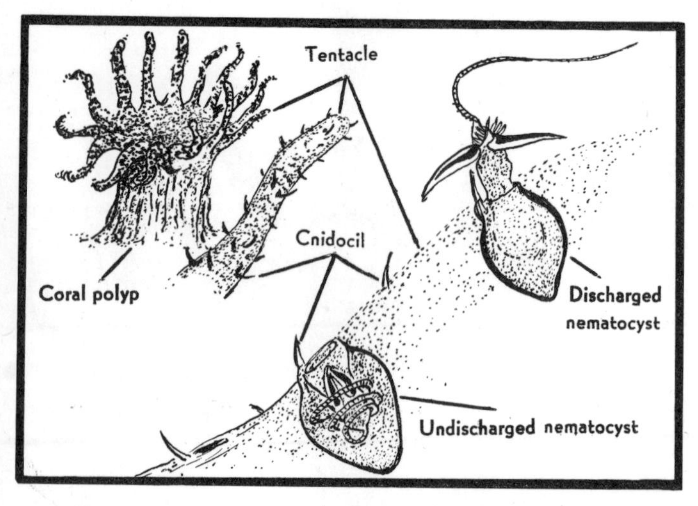

For many years it was believed that the nematocyst was activated by touch. Now it is generally agreed that substances exuded by the prey act as the stimulus causing the discharge. Acetic acid and methelene blue have been used in laboratory experiments to bring about a "firing" of the nematocysts. The cnidocil—once thought to act as a kind of mechanical trigger—is believed to be sensitive to such substances.

Some fish live unharmed within the confines of certain jellyfish and anemones. However if one of these fish is injured and its flesh exposed, it is liable to cause action on the part of the nematocysts and the fish will soon become food for its host.

The primary function of the tentacles, other than bearing the stinging cells, appears to be providing the animal with more surface with which to gather food. They also carry the food to the mouth of the coelenterate so are capable of simple muscle action. Because the thread-like tethers of the nematocist tend to keep the prey attached to the tentacle, there is no call for the latter to entwine the food in any way.

Once inside the coelenterate, food is broken down by powerful protien-splitting enzymes, and then is distributed throughout the organism by means of wandering cells.

Classes and forms of coelenterates

The Pacific Ocean has a wide variety of representative coelenterates ranging from large jellyfish (more than two feet in diameter with tentacles well over two feet in length) to the most delicate and minute of coral. They are found in the coldest of waters to the warmest, from relatively deep environments to the surface. Some of them, notably a number of reef corals, can lie exposed to the air for hours and still survive.

There are several areas in the field of coelenterate study that remain unexplored, and perhaps the most interesting of these is phosphorescence. Some jellyfish, for example, give off a brilliant display of light when stimulated by either mechanical, electrical or chemical means. Several corals are able to produce light and maintain illumination for extended periods of time.

Hydroids

Hydroids look like plants and are very often mistaken for seaweed. Close observation, however, reveals polyps lining the branches of the hydroid. These polyps are called hydranths. Food is taken into the system through the tentacles and mouths of the polyps, then through an elaborate system of canals in the stems of the hydroid as well as its branches. Like many other polyps in the phylum *Coelenterata*, the hydranth's tentacles carry stinging cells.

A curious feature of some hydroids is the variation of generations. These hydroids do not produce other hydroids; instead, free swimming, small jellyfish are born. In turn, these jellyfish produce, not other jellyfish, but hydroids. This is what scientists refer to as metagenesis.

The most common hydroid found in the shallow shore line areas of the Hawaiian Islands is the *Pennaria*. Its black stem and branches display their white hydranths to best advantage. *Pennaria* is found attached to stones or wharves where it may attain a height of about six inches.

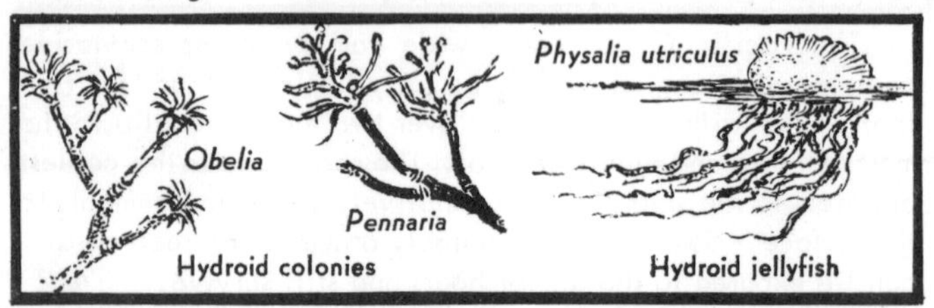

Hydroid colonies   Hydroid jellyfish

Hydroid corals (Hydrocorallina) are like regular corals in that they secrete a skeleton similar to that which builds reefs. The hydranths and other parts of the hydroid coral are very much like other hydroids. The beautiful red or pink *Stylaster* is an example of the Hydrocorallina, as is the *Millepora*, the skeleton of which is brown or yellow.

Jellyfish

There is a wide range of medusae (jellyfish) in Pacific waters. Two species, *Charybdea moseri* and *Solmaris insculpta*, are believed to be peculiar to Hawaii although representatives may be found elsewhere in closely allied species.

Perhaps the most familiar jellyfish to Islanders is the Portuguese man-of-war, *Physalia utriculus*. It is best identified by its blue or purple "sail" or float. *Physalia* has long tentacles, some of which are armed with nematocysts which sting and paralyze its prey, and this ability to deliver such a toxic and painful sting has made the Portuguese man-of-war a much respected animal. The gas filled "sail" measures from 3 to 12 inches in length and the feeding tentacles may extend up to 30 feet.

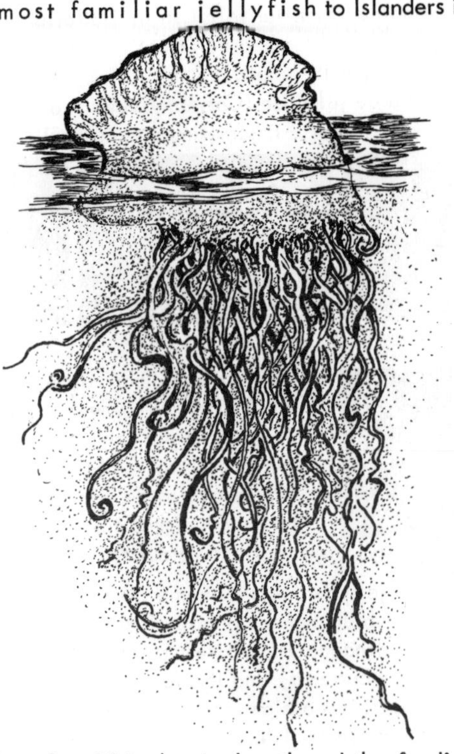

Almost as feared as the Portuguese man-of-war, is the bell-shaped jellyfish, ***Charybdea moseri***. It has but 4 tentacles -- yet they are dangerous enough to seriously injure a swimmer. Curiously enough, little fishes take refuge among the tentacles and are never hurt. The bell may be up to 4 inches deep and two inches wide -- though one with a bell twelve inches across was found in Pearl Harbor.

Another familiar jellyfish in Hawaiian waters is the ***Velella pacifica***. Its bluish-green float is flat and generally measures 2 or 3 inches in length.

A large jellyfish (***Cotylorhyzoides pacifica***) is often found in Pearl Harbor. It sometimes exceeds 12 inches in diameter.

Charybdea moseri

Solmaris insculpta

Velella pacifica

Atolla alexandri

Pelagia panopyra

Sea Wasp

A curious and dangerous jellyfish about which little is known has caused authorities in Australia great concern. It is *Chironex fleckeri*, the Sea Wasp, and the venom in its tentacles may prove to be one of the strongest science has encountered to date. This jellyfish seems most active during summer months.

The jellyfish gets its name from the gelatinous mesoglea or "jelly" which makes up most of its body. It is a free-swimming animal although its movement depends a great deal on the wind and currents, and the feeble contractions of its bell.

Jellyfishes are generally found alone, but often they can be seen as parts of large schools. This does not suggest an instinct for social intercourse, merely the result of the winds and currents which have so much control over their existences.

Sea anemone

Anemones are marine polyps which look very much like flowers. Their size varies from very small to quite large... 3-7 inches in diameter. What distinguishes the anemone from other coelenterates is the gullet which leads from the mouth to the animal's stomach.

Anemones have been categorized by scientists Anthozoa and share this classification with the sea pens as well as stony corals. Anemones are found in warmer waters, although some are also found in colder climates.

The common sea anemone is very much like its close relative the coral polyp in many respects: Both have tentacles surrounding slit-like mouths. Both attach themselves at their base to solid objects. They feed alike and multiply alike, and both can withdraw to an almost flat surface when disturbed.

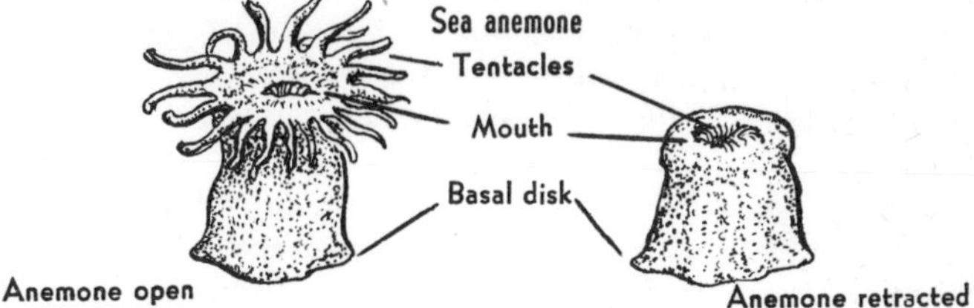

Unlike the coral polyp, the anemone is out and feeding most of the time. It does not build a lime cup as does the coral polyp. And, unlike the coral, which must live a sedentary life, it is believed that the sea anemone can creep slowly on its pedal disc. Its food is made up of fishes, mollusks, and crustacea which it catches by its tentacles and paralizes by nematocysts. Anemones are in turn eaten by fishes, larger crustacea, crabs, starfishes and even some mollusks.

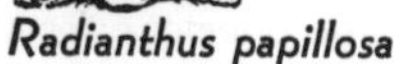

Radianthus papillosa

Sagartia pugnax

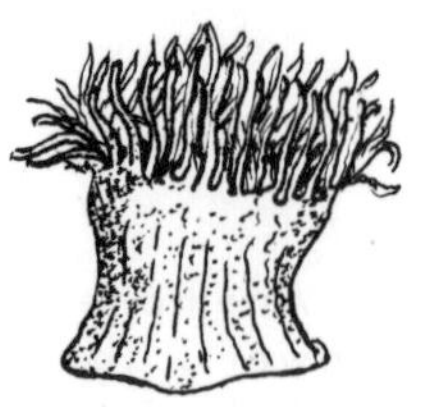

Anemonia mutabilis

The Pacific Ocean has many species of anemone, most of which are found in Hawaiian waters. One of the largest is *Radianthus papillosa* which grows to nearly a foot in length and possesses over 500 tentacles. It lives, curiously enough, almost completely buried in the silt and sand of shallow waters.

Another common species is *Anemonia mutabilas* which is small, reaching a height of only one and one-half inches and possessing about 300 tentacles.

An interesting anemone is *Nectothelia lilae* which lives attached to a plant and can swim like a jellyfish when freed.

Delicate *Sagartia pusilla* is also quite small. It has under 50 tentacles and, like a chameleon, can assume the color of its environment. One of the more familiar anemones is *Sagartia pugnax* which is pure white in color and is usually carried on the back of a crab. Many swimmers and divers are familiar with *Sagartia longa* which is found under stones in shallow water. It has short green and white tentacles, a soft body, and is light brown in color. *Phillia humilis* can also be found clinging to the underside of a stone in shallow water. It too has few tentacles, under twenty-five in number.

The *Cladactella manni* is a dark green and orange anemone about four inches in height and two and one-half inches in diameter. Its tentacles are long but unlike most other species it cannot retract.

Coral

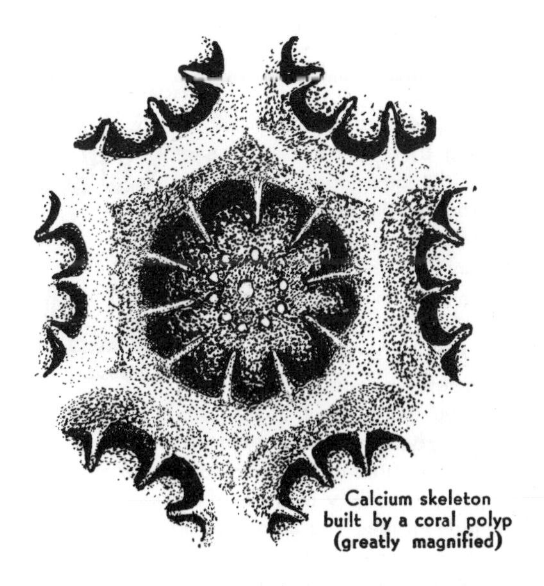

Calcium skeleton built by a coral polyp (greatly magnified)

A great deal of land on earth today exists because of animal life. Perhaps the most prolific builder is a creature so small that a magnifying glass or microscope must be used to study it closely. It remains in one place all its life, with the possible exception of the days following its birth. Yet, despite its size and sedentary existence, the *Coral polyp* is responsible for countless islands throughout the tropical world. It built The Great Barrier Reef, 1260 miles in length, along the north-eastern coast of Australia.

The coral polyp utilizes the rich chemical content that lies in solution in sea water. Just as the clam and snail utilize this resource to build their shells, and the lobster and crab use it to strengthen theirs, the reef building coral animal uses it to form the coral skeleton. It does this by taking in the calcium builders and secreting lime under the base and around the sides of its body. The polyp, in effect, builds a cup-like house which is rock hard and durable. Depending upon the species and the development of the polyp, the size of the cup housing an adult animal may range from 1/32" to a half inch in diameter and anywhere from twice to three or four times its diameter in depth.

THE CORAL POLYP

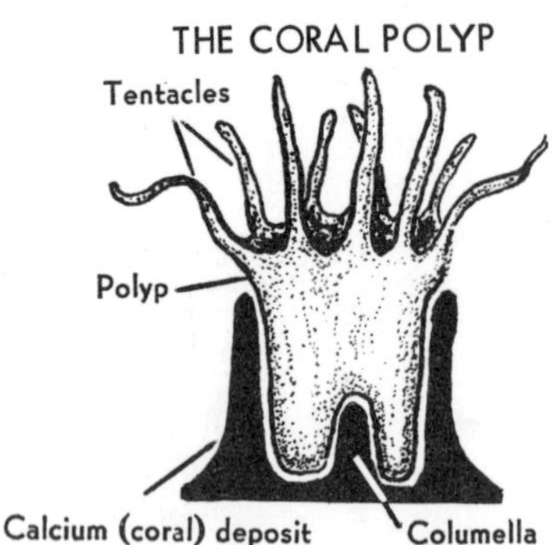

A coral head made up of 1/32" cups has 36 to 40 cups per square inch in a honey-combed, symmetrical pattern. This means, of course, 36 to 40 coral polyps created that much reef.

Coral polyps multiply in either of two ways, by budding or by free-swimming larvae. The coral bud grows out from the parent like a branch of a tree. This is the process which has resulted in the building of thousands of miles of reefs all over the tropical world. The free-swimming larvae, by contrast, are swept along with the currents. They make up part of the plankton of the sea. Those which survive the journey find a solid place to take up housebuilding (such as a rock, dead coral or other undersea formation). Observation has shown the duration of the free-swimming larval stage to be anywhere from 6 to 22 days, again depending on the species. Once the polyp (like its close relative the sea anemone) attaches itself to an object, it is difficult to remove.

Reef building corals can exist only in warm tropical waters where the temperature seldom goes lower than 68 or 70 degrees (F.). This distinguishes the reef-builder from its colder climate relatives such as *Balanophyllia elegans*, a solitary coral that is found as far north as Puget Sound in Washington State. Unlike many cold water species too, reef coral requires sunlight because of a relationship it has with algae. The coral polyp provides a home for an undetermined amount of algae and the arrangement is one of convenient mutual help, with the polyp supplying the algae carbon dioxide, and the algae, in turn, returning the favor by giving off oxygen which the polyp needs. This is known to scientists as symbiosis, wherein unlike organisms live together advantageously.

The polyp, like the sea anemone, sustains itself by catching with its tentacles minute creatures which come within reach. Plankton is the principal food of the animals. When something touches a tentacle, sensitive triggers cause hundreds of nematocysts to discharge barbed harpoons into the prey. The harpoon remains attached to the tentacle by a tether which ensures the food doesn't escape, so it is not necessary for the tentacle to encircle prey in order to draw it to the polyp's mouth.

Stony Corals

Porites compressa

Pocillopora damicornis

Porites lobata

Stony corals are the most common of the reef builders and are found in both shallow and deep water. Those which inhabit shoreline areas are generally more delicately formed, with slender, graceful branches. The representatives which grow farther out are more in keeping with a rougher environment, being solid, squat and thick.

The Pacific Ocean's coral world includes many families and a great diversity of forms. Over 120 species and varieties are known in Hawaiian waters alone.

Dendrophyllia manni

Acropora palmata

Meandrina meandrites

Black Coral

Black Coral is found in many of the world's seas, yet it still ranks among the most rare of corals. The reason for its rarity is the depths at which it develops and grows. Black Coral survives best in waters over 131 feet (40 meters). The fact that it can also develop in shallower waters, but only in light-protected areas such as caves and beneath overhangs, suggests that it is sensitive to light.

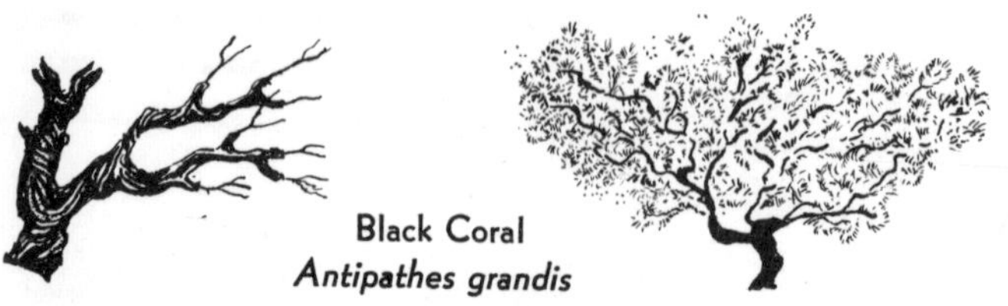

Black Coral
Antipathes grandis

The coral is difficult to obtain because of the depths required for its survival. Divers who mine it risk their lives. Black Coral's economic value is determined by its rarity and upon the fine jewelry and sculpture made from it. It is hard and takes a beautiful shine.

ECHINODERMS

Phylum *Echinodermata*

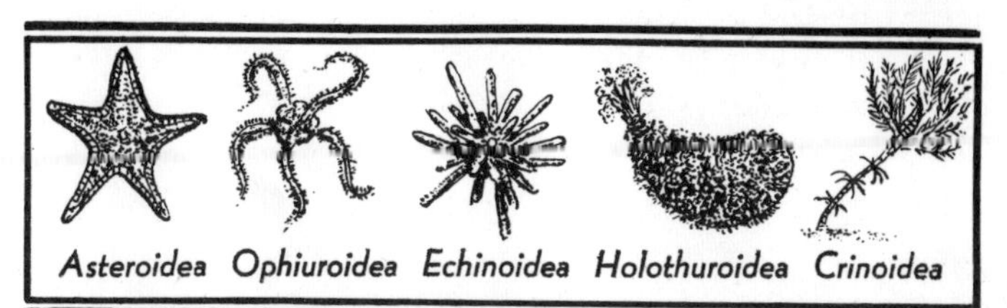

The echinoderms are a group of highly developed invertebrates. They are marine animals which, though very slow moving, are generally free-living and quite unique. They possess complex organs including a water vascular system and unusual powers of regeneration. Because of their radial symmetry, the echinoderms were once classified with the coelenterates. However, because of their calcerous skeleton, their water system and other characteristics, they have been placed in a separate and higher grouping.

Until the middle of the nineteenth century, for example, echinoderms were classed with coelenterates. Then it became recognized that,while coelenterates had a simple gut and a simple mouth, echinoderms had an alimentary canal with mouth and anus and the body organs were encased in a distinct body cavity.

Two features of echinoderms are noteworthy. First, the adults display a body pattern numerically structured in fives (pentamery). Second, all echinoderms possess tube feet. Only the sea cucumbers, several of the urchins, and a few of the starfishes depart from the pentamerous symmetry feature.

The echinoderms include the starfishes, sea urchins, sea cucumbers, and the sand dollars. The crinoids are of special interest because they are closest to the ancestral form.

Asteroidea: Starfishes

Starfishes are found close to shore and may be observed, during daylight hours, clinging to rocks or coral heads; they have rudimentary eyes on the tips of each ray, making them light sensitive. Like other echinoderms, starfishes have a taste for mollusks and frequently compete with fishermen for the oyster and clam beds.

Ophiuroidea: Brittle Stars and Basket Stars

The brittle star favors coral crevices or the undersides of rocks where it waits for food to float by. It appears to be a shy animal, to be seen only when one of its five arms darts out to snatch a piece of food. As its name suggests, the brittle star is prone to losing an arm or two, but, like all echinoderms, its regenerative powers are rapid.

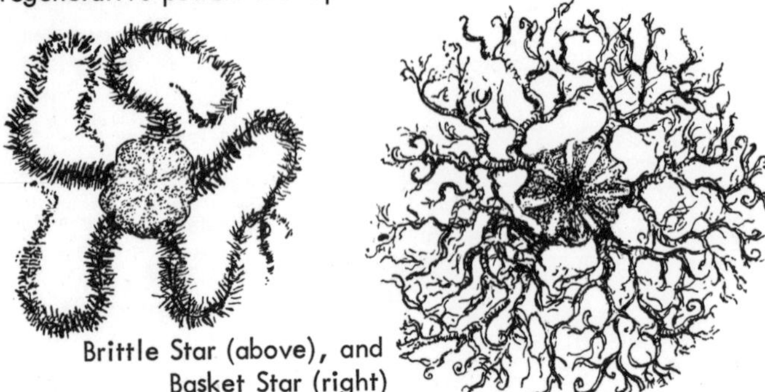

Brittle Star (above), and
Basket Star (right)

The basket star has multi-branched arms and closely resembles a plant or very small tree. It is a shallow water dweller and may often be seen at night "walking" along the bottom on the tips of its arms in search of food, especially crustaceans such as tiny shrimp. A large specimen may have arms up to twenty-four inches long and may have as many as 81,000 terminal branches.

Echinoidea: Sea Urchins, Sand Dollars and Heart Urchins

The spines of the sea urchin make it unique. While the animal uses its tube feet to move from place to place, it may also use these spines, which are constantly moving in circles apparently on the alert for food or attack. When so much as a shadow falls on the urchin, a number of spines immediately point to the source of the shadow. These spines, which in some urchins can be as long as 10 or 12 inches, can result in a painful sting, for they are barbed at the tip and can penetrate the skin very easily. While the wound is not usually fatal to man, it can be unpleasant; antibiotics usually have no effect and if the tip of a spine breaks off in the wound, only the body's natural defenses can dissolve the particles and venom.

There are about 700 species of urchins in the world but Hawaiian waters provide homes for only 20 of these, most of which are also spread across the Pacific and through the Indian Oceans.

The sand dollar is very seldom seen because it is a burrowing animal that prefers hiding beneath the sand rather than in crevices or under rocks. Like the sea urchin, the sand dollar is covered with spines but these are short and give the sensation of soft fur when rubbed. Diatoms and algae form the major part of the sand dollar's diet, but it will accept almost any tiny particle of food.

Holothuroidea: Sea Cucumbers

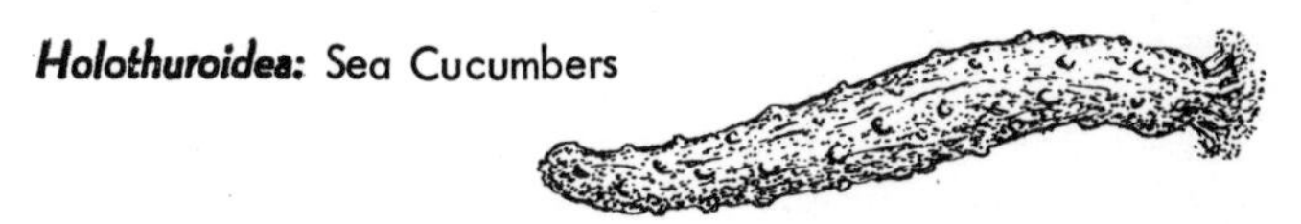

Nearly two thousand years ago a Roman, Pliny the Elder, was fascinated by a sausage shaped creature of the sea and called it a "sea cucumber". The name has remained with this echinoderm to the present time.

There are a number of unique features which separate these animals from other echinoderms. Sea Cucumbers, for example, depart from the pentagonal design of their relatives and are able to throw out most of their internal organs when disturbed, regrowing them within two months. In addition, some species of sea cucumbers emit a poison which can kill fish; this poison (called Holothurin) is being analyzed for its potential as a helpful drug for mankind.

In appearance, the sea cucumber is flabby and unattractive. It inhabits the ocean's bottom, taking in mud, silt and sand through a cluster of tentacles, filtering out the food and expelling the waste. The ecological value of this animal is two fold. Primarily, it helps to keep the ocean floor clean by acting as a vacuum cleaner but it also serves as a source of food, particularly in the Orient. The basis of the recipes for "trepang", "beche-de-mer" and several soup dishes is the sea cucumber's body wall.

Although sea cucumbers prefer colder water areas, their distribution is worldwide, particularly in tropical seas. There are over twenty species in Hawaiian waters, representatives of which inhabit most of the tropical Pacific and Indian Oceans. The Hawaiian sea cucumbers are, however, generally longer and thinner than their colder water cousins; specimens over forty inches in length have been found on occasion, but the average size is from ten to eighteen inches in length and from three to six inches around.

Crinoidea: Sea Lilies and Feather Stars

The Indo-Pacific region is rich in the number of sea lilies and "feather stars" (crinoids) inhabiting the depths. This class of echinoderm is plant-like in appearance and is considered among the more attractive of sea animals. Crinoids are also among the sea's oldest inhabitants; until recently they were known only through about 5,000 fossil forms, but deep sea excavations have proved their continuing existence.

Crown of Thorns

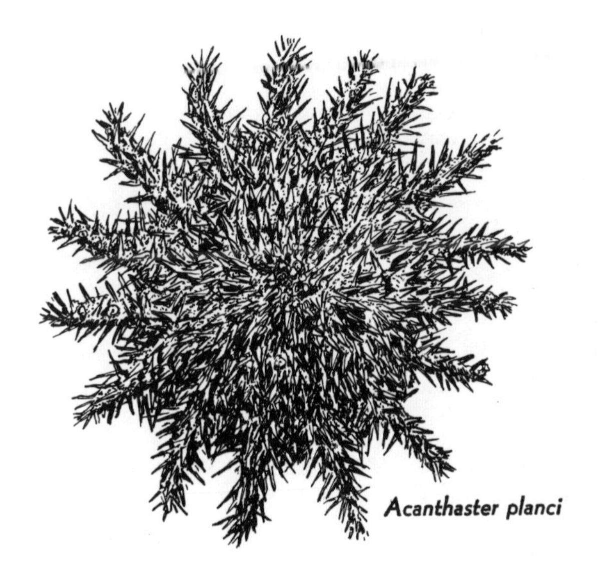

Acanthaster planci

The Crown of Thorns Starfish, known as *Acanthaster planci* to scientists, came into prominence in 1969 when it was discovered to be devastating reefs in a number of areas around the South Pacific. In the High Islands of the Marianas, particularly Guam, it is reported to have killed 90% of the reef in parts of the islands. Programs were begun which were designed to find means of combatting the "invasion".

Problem

It has been difficult to learn why this starfish should suddenly become so damaging. Many scientists feel that there has been an upset in the ecological balance of certain Pacific areas that allowed a proliferation of the starfish. Others are not as concerned; they feel the imbalance is temporary.

It is unfortunate that the animal's natural foe, the Triton, has been reduced in numbers by shell collectors as it appears to be the one creature best able to maintain the proper balance. However, large Hermit crabs have been observed attacking and destroying ***Acanthaster***, as has the rare shrimp ***Hymenocera elegans***. Experiments were underway in 1970 designed to measure the value of establishing elegans in affected Australian reef areas. Naturally, a cautious eye was being kept on any ecological problems that might arise. If the shrimp was successful in controlling the starfish and, in turn, multiplied to a point of imbalance, scientists might well be faced with another menace.

Activity

Observations have shown that the ***Acanthaster*** moves upon a coral head and, extruding its stomach, an ability common to starfish, devours the thousands of tiny polyps which build the coral and, subsequently, the reef. The belief has been expressed that ***Acanthaster*** also exudes its digestive juices in the process. If this is the case, it's no wonder that even the polyps which the starfish cannot possibly reach—those deep within the complex canyons of a coral head—are also killed.

What remains after the ***Acanthaster*** moves on is the clean calcium skeleton of a once thriving coral colony. Scientists worry that the subsequent action of waves and currents will ultimately erode the reefs built up over the thousands of years by the generations of coral polyps. Where these reefs protect islands from pounding surf, the islands would suffer.

Habits

Like most starfish, *Acanthaster* is a kind of bottom scavenger — a most necessary creature in cleaning up after its fellow inhabitants of the sea. Its larvae are released into the currents to develop and be carried along to whatever destination suits the element. In effect, it is very much like plankton. Ironically, the larvae of starfish make up part of the diet of the coral polyp.

Description

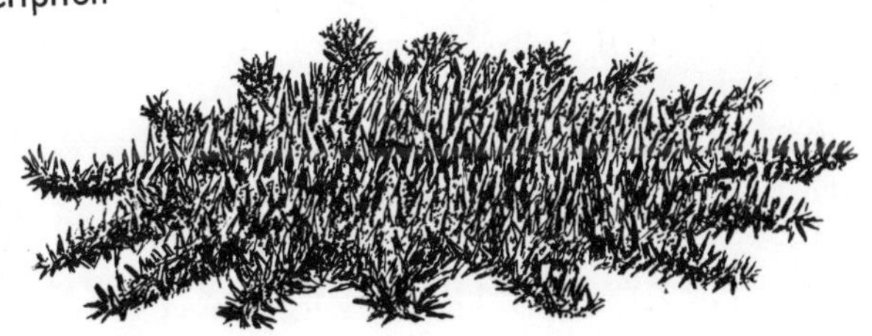

Acanthaster is known to grow up to 2 feet across. Its back is covered by sharp spines (thorns) which emit toxicity. The starfish generally has 16 rays and these rays are short in comparison to the rest of its body. The creature moves about by means of hundreds of tube-like feet, each tipped by a tiny suction cup. The Crown of Thorns' multi-colored back is an attractive blending of black, brown, red, orange, blue and grey.

Like many echinoderms the Crown of Thorns can regenerate a lost ray, and cutting one in half results in two starfish.

Its distribution is wide; from the Red Sea to the Philippines and Ryukyu Islands, Molucca Islands, Fiji, Samoa, Hawaii, etc. The Crown of Thorns is known to have been in Hawaiian waters for many years but has never been considered dangerous.

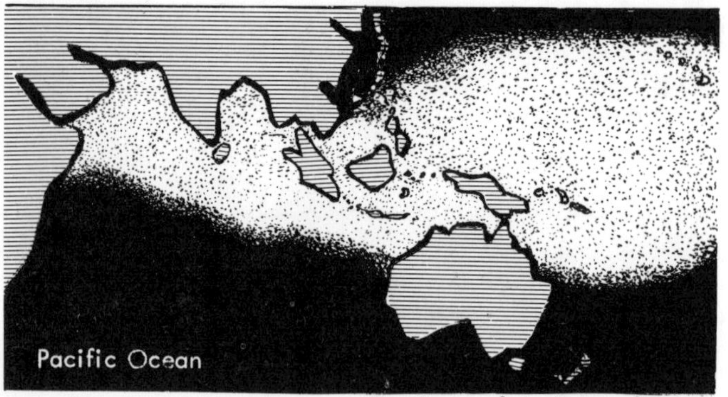

Distribution of the Crown of Thorns

MOLLUSKS

Phylum *Mollusca*

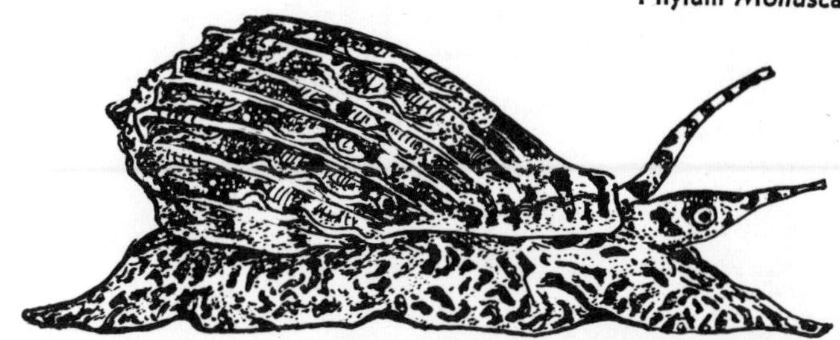

Mollusks comprise one of the largest animal groups in the world. Various estimates have placed the number of species on the face of the earth today at between 60,000 and 100,000.

There is no standard "form" which can specifically identify the mollusk. They are bilateral in shape, not segmented, and they extend from front to back without partitions.

Scientists have classified mollusks into five main groups: *

1) Amphineura (double-nerve). The chitons.

2) Gastropoda (stomach-footed). The snails, slugs, and related animals. This is the largest group. About 75% of all mollusks are gastropoda.

3) Scaphopoda (boot-footed). The tooth or tusk shell.

4) Pelecypoda (hatchet-footed). Clams, mussels and oysters.

5) Cephalopoda (head-footed). Octopods, squids and nautiloids. These are the largest of the mollusks.

The shells illustrated on the following pages begin with the gastropods, followed by the chiton, the tusk and the clam shells. Finally, the cephalopoda are depicted.

*However, the recent addition of the primitive *Monoplacophora* has meant that a sixth group is now generally included.

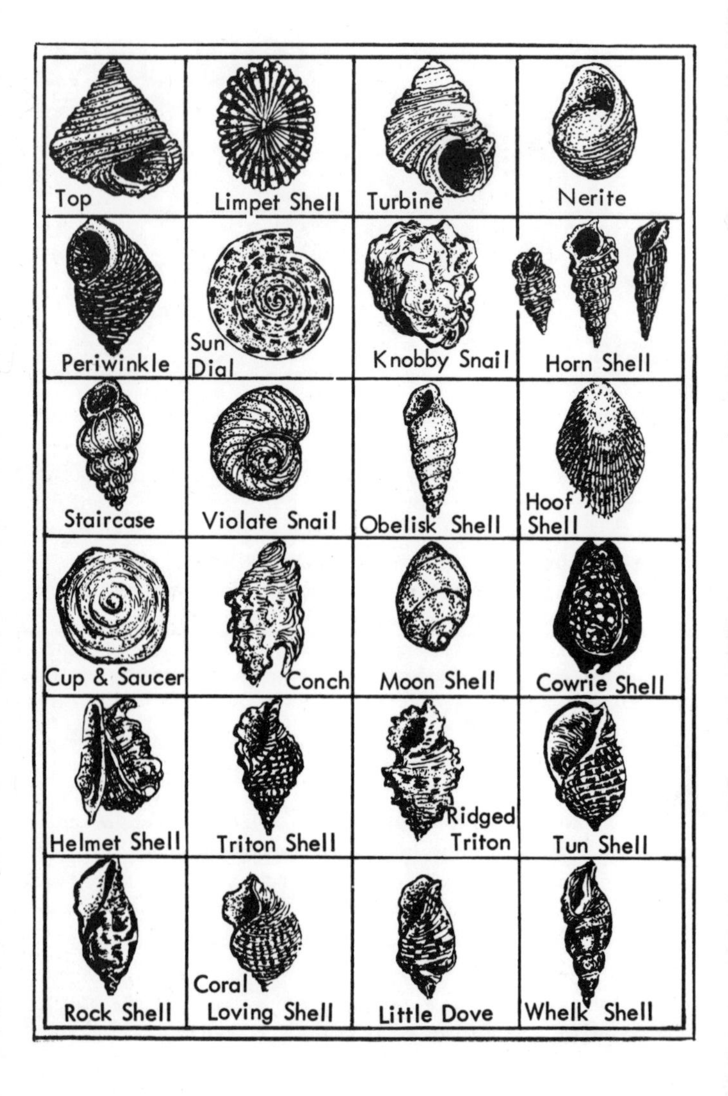
Top
Limpet Shell
Turbine
Nerite
Periwinkle
Sun Dial
Knobby Snail
Horn Shell
Staircase
Violate Snail
Obelisk Shell
Hoof Shell
Cup & Saucer
Conch
Moon Shell
Cowrie Shell
Helmet Shell
Triton Shell
Ridged Triton
Tun Shell
Rock Shell
Coral Loving Shell
Little Dove
Whelk Shell

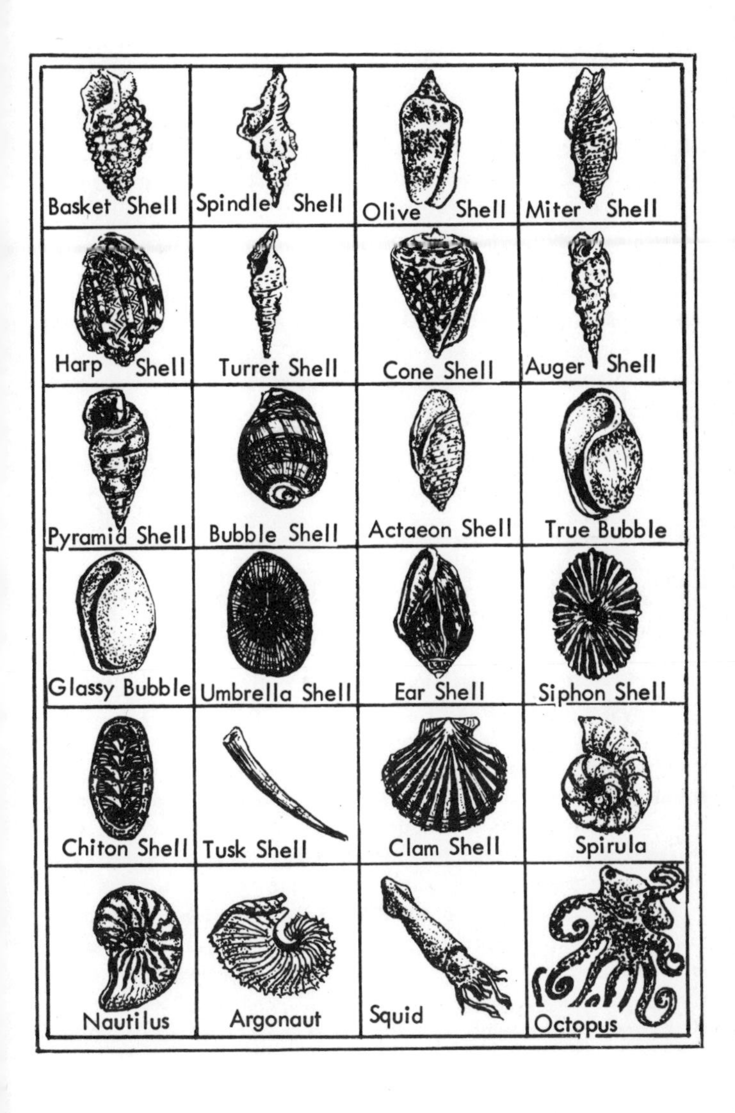
Basket Shell
Spindle Shell
Olive Shell
Miter Shell
Harp Shell
Turret Shell
Cone Shell
Auger Shell
Pyramid Shell
Bubble Shell
Actaeon Shell
True Bubble
Glassy Bubble
Umbrella Shell
Ear Shell
Siphon Shell
Chiton Shell
Tusk Shell
Clam Shell
Spirula
Nautilus
Argonaut
Squid
Octopus

Nudibranchs

Nudibranchs are beautiful multi-colored sea slugs which are generally found in shallower waters. The animals are shell-less mollusks which have neither mantle cavity nor true gills. Nudibranchs breathe through a complex of secondary gills or through the body surface.

The chief diet of many nudibranchs is coelenterates (i.e., anemones and hydroids) and sometimes other mollusks. It is believed that the nudibranchs have the unique ability to avoid the ill effects of the nematocysts of the anemones and, in fact, use the stinging cells to their own advantage after digesting them.

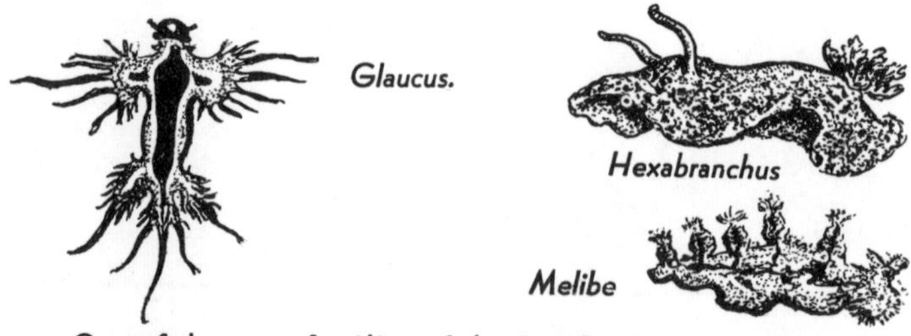

Glaucus.

Hexabranchus

Melibe

One of the more familiar of the Pacific Ocean nudibranchs is Glaucus. It is pelagic (found in the open ocean) but it is often seen along shoreline areas when the winds sweep it toward and onto the beach. Glaucus is a surface dweller kept afloat by means of gas-filled chambers within it body. This nudibranch feeds on the deadly Portuguses man-o-war (see page 13) and stores the nematocysts of the jellyfish. While Glaucus is normally harmless to humans there are cases on record of Australian bathers being stung by it. Glaucus had joined its cousins in demolishing many of the jellyfishes which had infested a beach area. The stings suffered by the bathers were similar in almost all respects to those generally administered by the Portuguese man-o-war. Glaucus seldom exceeds 2" in length.

Shells

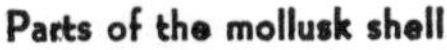

Parts of the mollusk shell

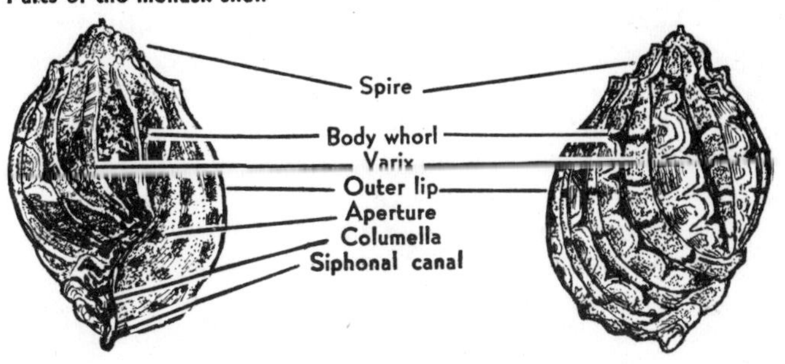

Composition

The mollusk shell is composed of calcium carbonate crystals, the end result of a process which begins with salt contained in the animal's blood in liquid form. This calcium is concentrated through osmosis by the mantle and deposited in layers to form the shell. There are, in many cases, several types of these layers, each one built by a different part of the mantle. The crystallization process results in either a calcite or an argonite layer. The finished product's cross section displays a structure of laminations which of course tend to give strength and rigidity to an otherwise brittle shell.

The interesting, often beautiful, color patterns which make the shell so attractive are caused by pigment cells located throughout the mollusk-animal's body, particularly the mantle. Various pigments, obtained for the most part through the food consumed early in the animal's life, are concentrated by the cells and, in turn, these cells tint the newly forming shell. The individual designs depend upon the movements of "wandering" cells.

The Largest Shell

The Giant Clam is found in such areas of the Pacific as the Great Barrier Reef of Australia, Eniwetok and Kwajalein. In some parts of the Pacific it is called the "giant killer clam" because it has reportedly clamped shut on the arms or legs of unwary divers, holding them until they drown. While the mollusk is indeed capable of doing this, it is unlikely that there have been many fatalities of this sort because of the slow movement evident when the animal closes its shell.

This mollusk grows to weigh up to 500 lbs., and may have a shell nearly 5 feet wide.

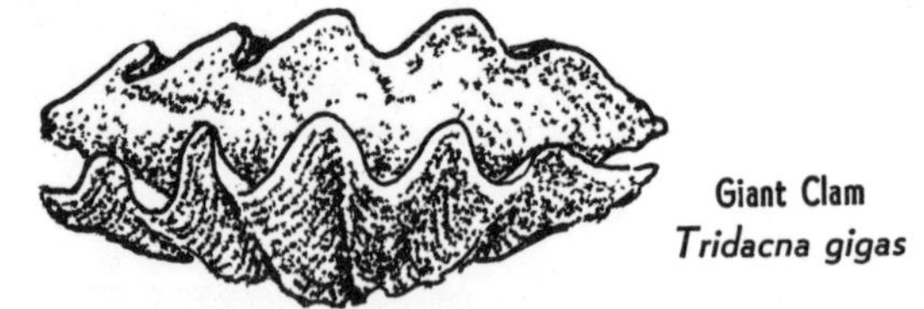

Giant Clam
Tridacna gigas

While shells like the Giant Clam, the Triton, Strombus and Helmet make good display items because of their size, and while other mollusks are admired for the lovely and colorful patterns which tint their outer surfaces, scientists and laymen alike are becoming interested in the beautiful interior structure of shells as well. With the possible exception of the coral polyp, no builder on earth, including man, can quite equal the exquisite patterns that trace the various growth stages of the mollusk.

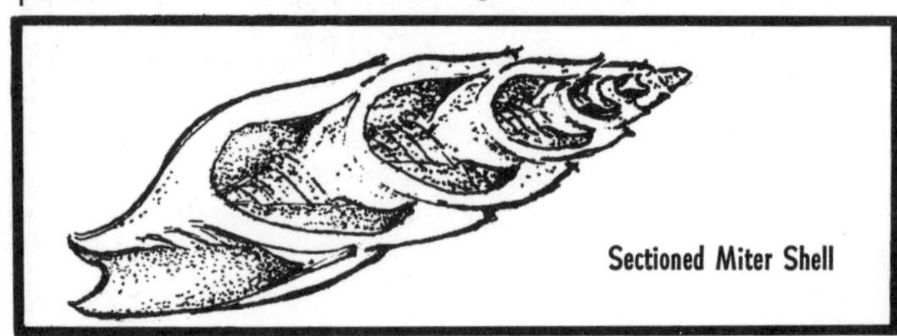

Sectioned Miter Shell

Sectioned shell of the
CHAMBERED NAUTILUS
Also see pages 39-40

Value

The value of shells fluctuates from decade to decade, more in keeping with prevailing economic trends than in the availability of individual specimens. This is not wholly true of course with such shells as the Golden Cowrie and Glory-of-the-sea, ***Conus gloria-maris***.

Because of its rarity, Glory-of-the-sea may command a price of up to $2000 and possibly more. What few are found are generally located in the southwest Pacific (Philippines, Solomons, etc.). There are only about 30 known specimens in private and museum collections around the world.

Golden Cowrie, by contrast, appears to be more valuable because of its demand than its rarity. Nevertheless, it remains one of the few shells with a high price tag of from $150 to $175 each. Another cowrie, a Caroline Island specimen, brought over $1300.

The Collection

Housing a collection does not have to be an expensive proposition. Cigar boxes are preferred by many hobbyists to store their shells, and records are kept on standard 3x5 cards. Perhaps the best data to keep as record are the location of the find, the date, and a bit of information concerning the habitat.

Scope

There are few branches of science that have been embraced by laymen on a scale displayed by conchology. Hundreds of thousands of people are actively involved in shell collecting as a hobby.

For anybody who might be interested in developing the few shells they possess into a more sophisticated collection, there are a number of up-to-date publications available. A list of these publications may be obtained on request from The U. S. National Museum, Division of Mollusks, Washington D. C. 20560.

Cephalopods

Octopus

Cuttlefish

Squid

Nautilus

The most highly evolved mollusks are the cephalopods. The somewhat cumbersome foot in other mollusks has developed into the cephalopod's more utilitarian tentacles and arms. Perhaps the most remarkable advance, however, has been in the region of the head particularly the eyes. A number of cephalopods possess an intelligence and eyesight far superior to most other of the sea invertebrates.

Cephalopods are a diverse group of animals though their similarities far exceed their differences. All have arms or tentacles ranging in number from eight for the octopods (Octopus and Argonaut) thru ten for the decapods (Squid and Spirula) to well over eighty for the Chambered (or Pearly) Nautilus.

Of the cephalopods, there is only one which grows a shell and lives in it: The nautilus. The argonaut (also called the paper nautilus) develops a parchment-like egg case by means of a specialized pair of arms. Spirula grows its shell internally.

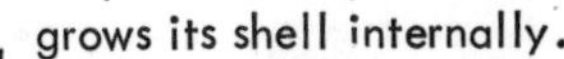

Argonaut and its egg-case

Spirula and, below, its shell

History

Cephalopods have inhabited the oceans since the Cambrian Period (600 to 500 million years ago). There are, today, well over 10,000 fossil specimens available which trace the history of this animal. Shells of the Giant Nautiloid, over 15 feet in length, date back 500 million years. The Cretaceous Period (135 to 65 million years ago) produced the ***Pachydiscus***, represented in museums now by a coiled shell over six and a half feet in diameter. Today, only the Chambered Nautilus remains a direct link to those giants of the past.

With but few exceptions, cephalopods are wide ranging. They are found in the coldest of climates or the warmest. Many are shallow water inhabitants while others wander the darkness of the deepest seas. Representatives of the octopus and squid families can be found in most environments. Only Spirula seems to prefer deep waters alone, while Nautilus appears to favor temperate areas and is found only in the Pacific Ocean.

Comparison

Cephalopods differ from other mollusks in several respects the most obvious of which is the location of the head in respect to the foot.

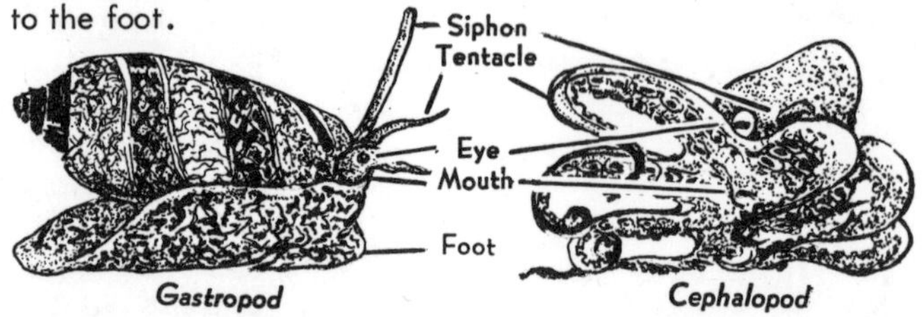

Gastropod *Cephalopod*

With the exception of the Nautilus, cephalopods are free of an encumbering shell. What they lack in protection they make up in mobility. While most mollusks can only move along the ocean bottom by means of a slow, undulating motion of muscles in their foot-slabs, the cephalopods, because the foot is actually a number of tentacles, can grasp and hold. While it is true that some mollusks secrete a mucus which helps in their movements over difficult obstacles, and others have numerous cilia which beat swiftly and so aid in locomotion, these are nevertheless poor substitutes for the multi-suckered tentacles of the cephalopod.

Some mollusks are believed able to use their siphon as an assist to movement. The octopus and squid have developed it into a superb aid to swimming. By forcing a jet of water out the siphon some squids have been able to attain speeds of up to 30 miles per hour. By being able to maneuver this appendage from side to side, the cephalopod is also able to control his movement's direction.

Intelligence

It is perhaps unfortunate that too little experimentation has been done on the intelligence of certain cephalopods, particularly the octopus. What limited work has taken place (notably in Naples, Italy) has shown the octopus capable of behavior which is very near true intelligence. It has been demonstrated that the octopus is able to learn and to remember. Also, it has long been known that octopus has an emotional range which covers the spectrum from fear to outrage.

As far as can be determined, little comparable study has been devoted to the squid or other cephalopods.

Nautilus

Several centuries ago mariners became quite familiar with a pair of creatures "sailing" along on the surface of the ocean. One of these animals had a pearly shell, the other a white, thin shell. They became known to seafarers as Nautilus or "little sailor". Eventually, one was designated Pearly Nautilus and the other, because of its paper-like shell, Paper Nautilus.

The Pearly Nautilus was rarely seen but the Paper Nautilus was very much in evidence in a number of ocean areas especially in the Mediterranean where they became known as Argonauta. This was an allusion to the Argonauts, the men who sailed with Jason to find the goldon fleece.

While the cephalopods were recognized for centuries, no one really knew what they were. As recently as the early 1800s it was believed that the Paper Nautilus had not grown its shell but had taken over the shell of some other mollusk much in the same way that a Hermit Crab might inhabit one. Finally, however, close study showed that the shell was indeed grown by the animal and, in fact, it was really no shell at all but an egg case. What was more, a baby animal was no older that 12 or 13 days when it began growing its egg case.

It was probably at about this stage of investigation that scientists of the day recognized that there was a difference between the Paper Nautilus and the Pearly Nautilus. Also, the latter became known as the Chambered Nautilus.

As suggested by its name, the Chambered Nautilus shell is made up of a series of "rooms", the last and largest of which houses the animal. Scientists have found by analysis that the chambers contain gases, rich in nitrogen, which enable Nautilus to control its bouyancy.

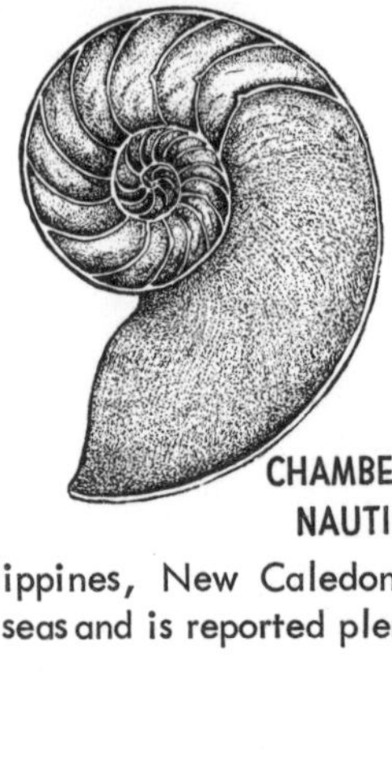

The "shell" of Argonaut, as already mentioned, is actually an egg case. It is thin but not fragile. The Nautilus shell, by contrast, is thicker and considerably stronger.

CHAMBERED NAUTILUS

Nautilus favors waters surrounding the Fiji Islands, the Philippines, New Caledonia, etc. Argonaut also favors temperate seas and is reported plentiful in the Australian area.

Octopus & Squid

OCTOPUS

Perhaps the most advanced mollusks are the Octopus and the Squid. These animals are cephalopods --"head-footed", so called because the "feet", i.e., the tentacles, are joined to the head. Even though they are closely related, the octopus and the squid are distinctly different mollusks. The octopus has eight tentacles--- the squid has ten, two of which are longer than the rest.

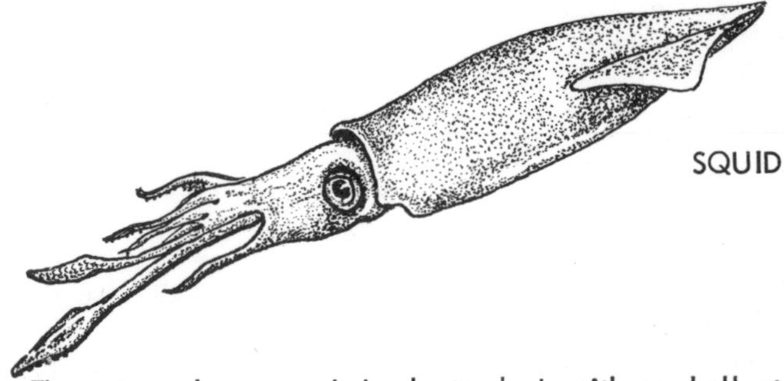

SQUID

The octopus has a rounded, shorter body with no shell at all. The squid's body is long and comparatively slender with fin-like lobes on each side. Beneath the mantle of some squid is a shell-like bone called, because of its shape, a "pen". This is the "cuttlebone" so familiar to people who keep birds as pets.

The octopus and squid are similar in many respects. They both have multi-suckered tentacles. Both propel themselves swiftly by means of an expelled jet of water. Both have ink sacs which enable the animal to squirt out a kind of watery smoke-screen when in danger. Both animals have well developed eyes.

Neither the squid nor the octopus grows very large in Hawaiian waters. An octopus with a 3-foot tentacle could be considered big. Squid in the Hawaiian chain seldom reach 3 feet from the top of their body to the tips of their long tentacles. Elsewhere, however, especially in colder waters, 9 or 10-foot tentacles on an octopus are not uncommon. As for squid, there are reports of animals up to 50 feet -- from mantle top to tentacle tip.

The octopus is considered by many scientists to be one of the more interesting cephalopods. It has the chameleon-like ability to change its color at will. By expanding or contracting thousands of tiny pigment cells in its skin, it can blend perfectly with its environment. These cells are of various hues: red, yellow, blue-green, etc. The octopus can contract or expand one set of these cells at one time, or all of them to produce a brilliant display of colors.

The female octopus lays from two dozen to 45,000 eggs and for about fifty days carefully watches over and protects them.* When hatched, a baby octopus is a perfect, though miniature, replica of the adult. There are well over one hundred twenty species of octopus all over the world. In the United States, this invertebrate is used primarily for animal food and fertilizer though gourmets prepare it for frying or for chowders. In Japan, it is served with soy sauce for hors d'oeuvres. In Italy, the octopus is sauteed in olive oil and fried. The Portuguese cook it in its inky fluid. One of the national dishes of Spain is octopus.

* Some octopods, however, release their eggs to the currents.

Squids are the fastest of the cephalopods, although some (mainly the cuttlefish, *Sepia*) are quite slow.

The long tentacles can be retracted or extended. It is these that the squid uses to trap its prey.

Like the octopus, the squid may be found in shallow or deep water. The giant squid is the favorite food of the sperm whale and stories of battles between the giants are legendary.

There are two species of octopus in Hawaiian waters. These are the day octopus, ***Octopus cyanea***, and the night octopus, ***Octopus ornatus***.

Of these two, the "day" is the most common and can be found in shallow water along reef areas during daylight hours. It is gray or brown in color and has a characteristic black spot below each eye; this spot is surrounded by a thin circle of brown. It is the larger of the two species and will have arms which reach 24 inches in length.

The "night" octopus is smaller and less abundant than the "day" octopus. It is reddish brown or orange in color and is marked by an interesting pattern of light dashes of color which extend in series along the dorsal side of the arms and onto the body. According to some experts, this octopus was more valuable than the day octopus. Both were hunted by torchlight at night and captured by hand or with nets or spears. Both species were used for food and occasionally to help in curing sickness which was caused by sorcery. If an octopus was held before the sick person, the illness might flee (*he'e*) from the body of the patient.

The octopus and the squid were well known in old Hawaii and were regarded as important and valuable animals. The bottom dwelling octopus was called *He'e* which means to flee or slide along. The more slender, free swimming squid was known as the *Mu he'e*. *Mu* here means changeable and unsteady and possibly refers to its backward and forward motion when poised at rest in the sea.

Perhaps because of its unearthly shape, the octopus has been described in literature as dangerous to man; a predator which crushes victims with its tentacles. Actually, the octopus is a retiring creature which will avoid humans if possible. True, there have been accounts of large octopus and giant squids attacking people, but such cases appear to be the exception not the rule.

Not to be taken lightly, however, is one octopus, identified scientifically as *Hapalochlaena maculosa,* popularly called the spotted octopus because of the blue circles spread over its entire body. It seldom exceeds four inches in size but it has been responsible for the deaths of a number of people. The danger posed by this creature is not in the tentacles but in its bite. The spotted octopus has salivary glands which contain enough venom to kill eight or ten men.

Blue ringed (or Spotted) Octopus
Hapalochlaena maculosa
(also *Octopus maculosis*)

Not much is known about this octopus yet. It seems to have enjoyed a population explosion particularly in Australian waters. It is known, however, how its bite affects human beings. Medical records attending the death of one victim describe how he suffered a bite on the neck. Several minutes later he complained of difficulty in swallowing and a dryness in the mouth. This was followed by stomach spasms which caused him to vomit. Muscular coördination suffered failure and he felt weak. He had difficulty speaking and breathing. Finally he couldn't breathe at all. No more than two hours after being bitten, he died.

Phylum *Arthropoda*

CRUSTACEANS

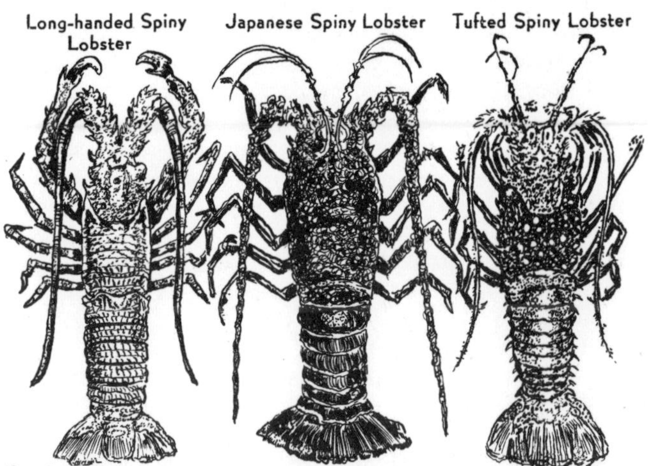

Crustacea

The class Crustacea is made up of crabs, lobsters, shrimps and their relatives and is numerically one of the largest classes of animals. Experts believe that the Crustacea have been inhabiting the earth for about 500 million years. Fossil remains have been found in rocks which date back to Paleozoic times. One of the unanswered questions about this class of animals concerns its ancestor. Some scientists believe that the phylum (*Arthropoda*) to which the Crustacea belong may have evolved from the annelid worms.

Crustaceans may inhabit either fresh or salt water and some of them are land dwellers. They breathe by means of gills which utilize the dissolved oxygen in water. In the case of the land animal, the gills are kept moist and the oxygen is drawn from the surrounding air.

One of the distinguishing features of all arthropods is an outer skeleton or shell which encloses each body. This shell is made of chitin and hardened by lime. The whole structure, rather than being formed of one piece, is divided into sections called somites. Somites are grouped to form regions of the animals' bodies: the head, thorax and abdomen.

The shell of the Crustacean prevents unrestricted growth. To escape this limitation, the animal has developed a means of shedding its shell and quickly forming a new one. When doing this, the Crustacean will find a protected place, remove some of the lime from the skeleton, shrink its body somewhat and then back out of the shell through a crack on the dorsal side. In many instances, large male crabs have been seen protecting females which have shed their shells. Such protection is necessary because the outer layer of the animal in this condition is quite soft and susceptible to attack.

One of the more important animals of the sea is also one of the smallest, the Copepod. This crustacean, no larger than a pinhead, is one of the prime sources of food for both small and large animals ranging from the lowly, sedentary coral polyp to the biggest of the magnificent far-ranging whales. The copepod is certainly one of the staples of marine life.

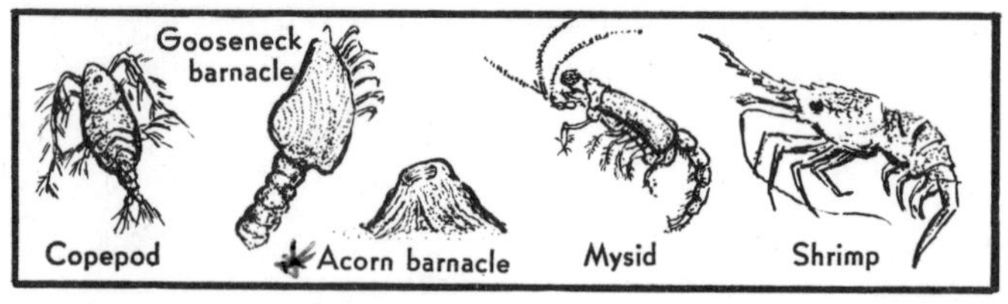

A big problem facing boat owners is a crustacean that lives a sedentary existence attached to an object such as a wharf piling, a rock or a boat. It is the barnacle, and, much like the coral polyp or anemone, it must rely on food coming within reach. One of the more curious of these animals is the Gooseneck Barnacle which in ancient times was believed to give birth to geese.

Mysids are tiny crustaceans which can be seen by the thousands at certain times of the year along shore and reef areas.

Shrimps, together with crabs and lobsters, are much sought as food for man.

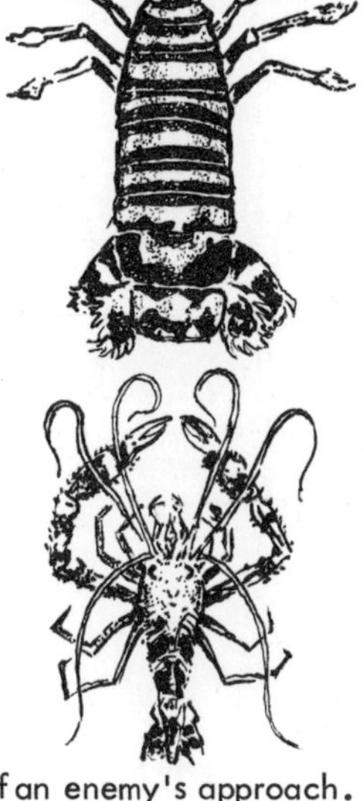

The Banded Squilla, also referred to as the Banded Mantis Shrimp, is a familiar animal throughout the tropical Pacific. It is found in the area which extends from Hawaii to the east coast of Africa. Because it lives along shorelines and in muddy bottom locations, this mantis shrimp is often caught in crab nets. It is an edible animal that has been reported to reach a length of eleven inches. It is called "mantis" because the shape of its claws, and the manner in which they are held, resemble those of a Praying Mantis.

Another crustacean that is prevalent throughout tropical areas is the attractively colored Spiny Prawn. It is predominantly red and white with six long, graceful and pure white feelers extending from its head. It is a gentle, shy animal which prefers shallow, calm water, and it limits its wanderings to areas where there are ample recesses into which it can quickly escape in case of an enemy's approach. This prawn seldom exceeds two inches in body length.

Lobsters

Most tropical Pacific lobster lack the large front claws that distinguish their Atlantic cousins. The Long-handed Spiny Lobster does have pinchers though they are very small in comparison to those of the American Lobster, for example. The rare Western Lobster appears to be the lone exception. Its pinchers are quite large and heavy and it is considered the only true lobster to be found in Hawaiian waters. However, even it is small by comparison, with a body seldom exceeding eight inches in length.

Western Lobster

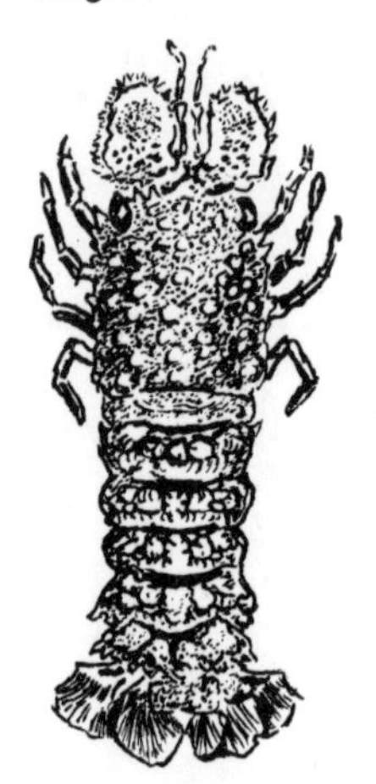

Regal Slipper Lobster

Slipper Lobsters are nocturnal in habit and they seek out remote and craggy areas of coral reefs where they establish hideouts. It is for these reasons that this crustacean is seldom seen. Some of them, like the Timid Slipper Lobster and the Regal Slipper Lobster are considered quite rare throughout the Pacific. The Regal Lobster in fact is believed to be an inhabitant of Hawaiian waters alone. Most of the various Slipper Lobsters are quite large and edible; the Scaly Slipper Lobster reaches a length of twelve or more inches. An exception is the Antarctic Slipper Lobster which is considered much too small (eight inches in length at most, and flat) to be economically valuable.

Crabs

The Pacific World's crab population is immense and varied, ranging from the curious Sea Cucumber Crab, which lives a symbiotic existence within the mouth and among the tentacles of sea cucumbers, to the Coconut Crab, a hermit crab which no longer lives in the water nor requires a shell to protect its once soft tail. In between are such interesting relatives as the Three-Toothed Frog Crab, named after the three spines or teeth located on each pincer, the Box Crab and the Hairy Anemone Crab.

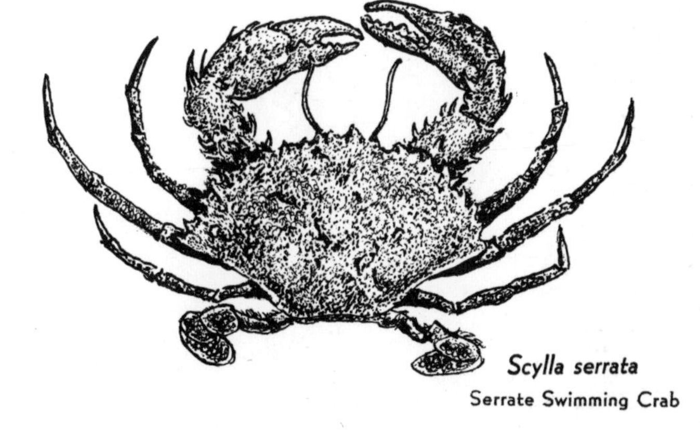

Scylla serrata
Serrate Swimming Crab

Among the more familiar crabs, perhaps the best known is the Serrate Swimming Crab, known in Hawaii as the "Samoan Crab". It is a large crab and considered excellent as food. The beautifully marked Blood-Spotted Swimming Crab is also well known in the tropical Pacific, and it, too, is caught and sold for food. The Long-Eyed Swimming Crab, so named because of the long stalks which support its eyes, is generally found close to shore. It is an edible species which is usually netted in large numbers in brackish water.

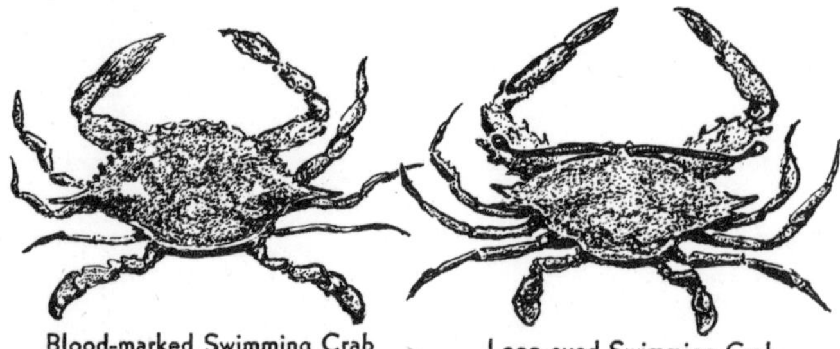

Blood-marked Swimming Crab — Long-eyed Swimming Crab

One of the more popular of the Hawaiian Island crustacea is the Red Frog Crab, perhaps better known as the Kona Crab. It is a burrowing animal that has a body and legs adapted to digging itself backward into the sand. This crab is considered especially delicious and is often served in restaurants or sold in markets.

It occurs from Hawaii to the coast of Africa and can be found at varying depths which range from thirty feet to over one hundred and fifty feet.

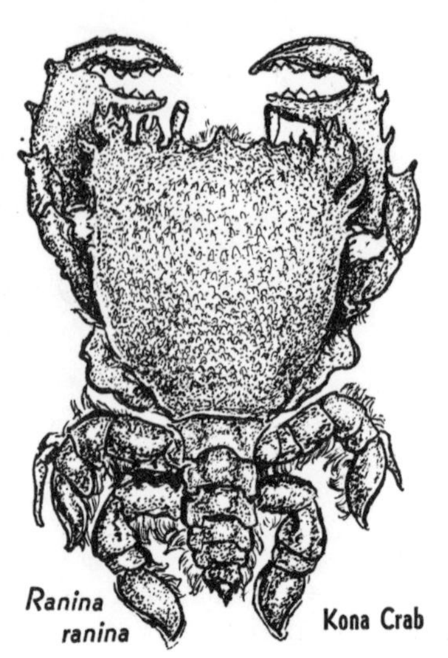

Ranina ranina — Kona Crab

Hermit crabs are among the more interesting animals of the sea. They are unlike other crabs in that their abdomens never develop hard shells. Because of this, they must seek out a shelter usually in the form of a mollusk shell, although almost all hollow objects will do as long as the crabs can back into them and carry them around as kinds of portable homes. This need for shelter causes a great deal of competition among hermit crabs for shells, and, as these animals grow, they are forced to seek larger shells.

Some hermit crabs decorate their adopted homes with sea anemone usually for protection but also for camouflage. When transferring from one shell to another, they can be seen gently removing the anemone from the abandoned to the new shell.

One of the more numerous of the tropical Pacific Hermits is the large Red Hermit crab. It is a deep water species seldom caught at depths less than fifty feet. This crab is found from Hawaii southward and to the west as far as the Indian Ocean.

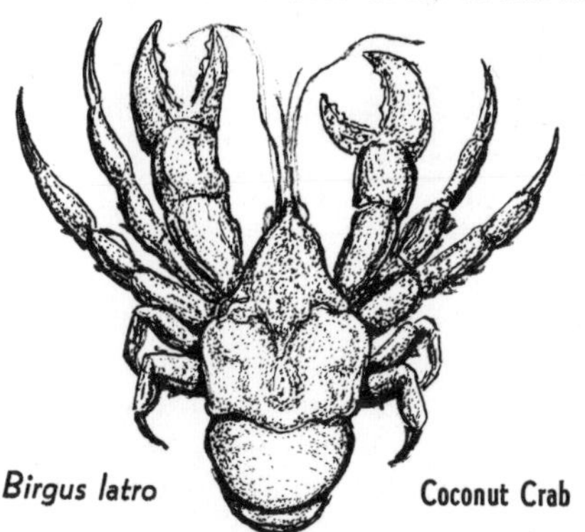

Birgus latro **Coconut Crab**

One of the largest land crustaceans is also a hermit crab: the Coconut Crab, which long ago adapted to conditions on land negating the need for a protective mollusk shell. It has powerful pincers and is reputedly able to climb coconut trees, snip off the nut, retrieve it on the ground and commence to husk and consume it. Proof of this appears lacking however.

This crab is not found in Hawaii but does occur in Palmyra and areas extending westward across the tropical Pacific to the Indian Ocean.

Beach goers are familiar with a swift moving little fellow which races in a tangential manner across the sand and disappears into tiny holes. This is the Ghost crab, two species of which are especially notable: the Horn-Eyed Ghost Crab, named for the horn at the end of each eyestalk, and the Telescope-Eyed Ghost Crab which has a pair of very long eyestalks. Both are widely distributed from Hawaii westward to the Red Sea.

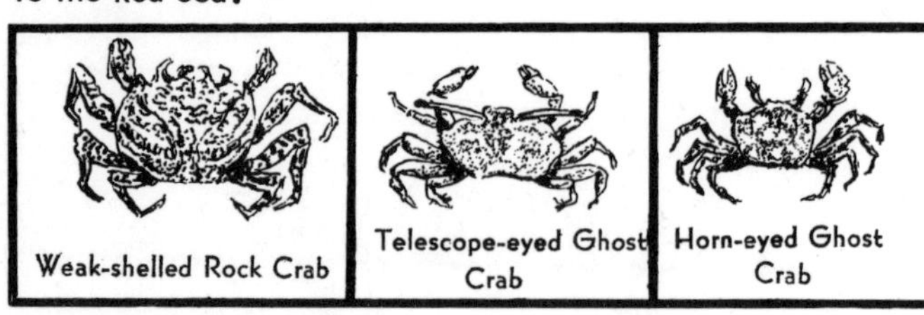
Weak-shelled Rock Crab | Telescope-eyed Ghost Crab | Horn-eyed Ghost Crab

There are a number of Pacific Ocean crabs which inhabit depths over 50 feet and, so, are not as familiar as those found in shallower waters. One of these is the Hawaiian Swimming crab, a small individual that appears to favor the outer side of reefs. The Red-Legged Swimming crab lives in waters beyond the reef areas. Its distribution is wide, extending from Hawaii westward to Japan and across the Indian Ocean to the Red Sea.

The Long-Spined Parthenopid Crab is a deep water species that is caught at about one hundred feet. It too enjoys a wide distribution, one that covers an area from Hawaii to the east coast of Africa.

Parthenopid Crab
Lambrus longispinis

Parthenopid crabs have a kind of natural camouflage to protect them from predators; they resemble coral or rocks. They have nothing to fear from humans either as they are poisonous. Ancient Hawaiians, recognizing this, placed a kapu (taboo) on them.

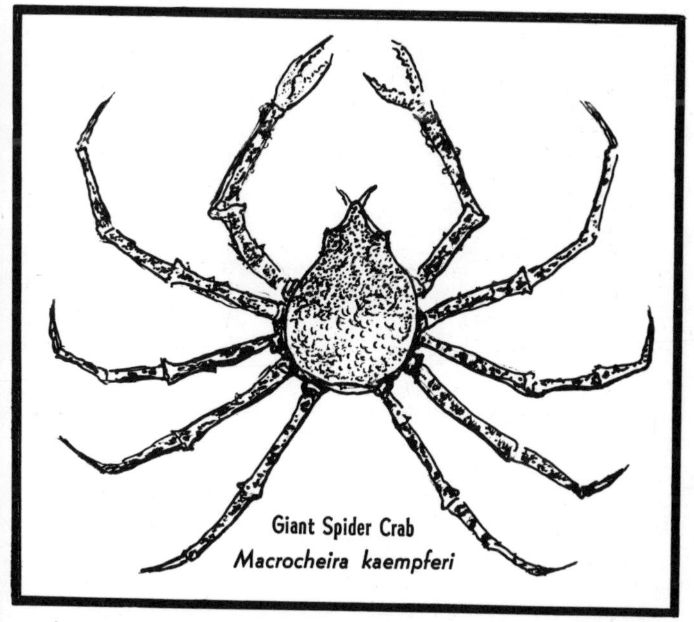

Giant Spider Crab
Macrocheira kaempferi

The earth's largest crustacean is the Spider crab found in deep waters off the coast of Japan. This arthropod, about which little is known, has long legs that can span 6 to 10 feet, and a carapace over a foot wide in adult form. It is not an edible crab nor is it found in abundance.

King Crab

The king crab, found in the northern Pacific from Alaska to Korea and Japan, is one of the largest crustaceans reaching a span of better than 5 feet and weighing over 23 pounds. That it is also one of the most delicious of the ocean crabs has made this arthropod the base of a multi-million dollar industry.

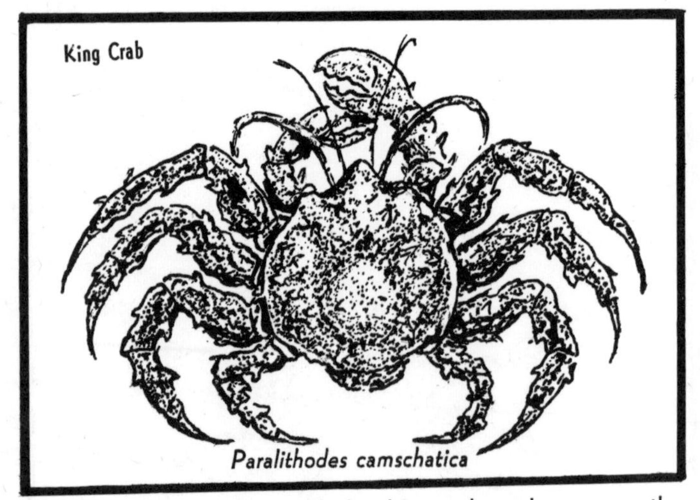

Paralithodes camschatica

Canners were processing Alaskan king crabs as long ago as the 1890's; the Japanese had developed catching techniques and were marketing the meat to a receptive world. Since World War II, American boats have been heavily engaged in fishing the crab. The trapping of this animal is primarily a winter business and the conditions in the far north during cold winter months are inhospitable. Nevertheless, heavy catch rates made possible by large crab pots capable of holding some half ton of crabs, and modern hauling gear, made regulations necessary. Today only males may be caught, and only those of an estimated age of 7 years or older.

In the past few years scientists have been devoting more time to the study of king crabs than ever before. One of the more interesting findings concerns juveniles and the manner in which they gather together, stacked atop each other in huge pods as they move along the ocean floor. The young crustaceans, up to 5 inches across, use this social pod as a means of protection against predators which are likely discouraged from attacking such a formidable mass.

Two other Pacific crustaceans rival the Giant spider crab and the King crab in size and market popularity respectively, the huge Tasmanian crab (not illustrated) and the Dungeness crab.

The legs of the Tasmanian crab, ***Pseudocarcinus gigas,*** reportedly span well over 6 feet. This crustacean is found in colder water areas of the South Pacific, notably Tasmania and Southern Australia, and has not been reported near the Great Barrier Reef.

Dungeness crab

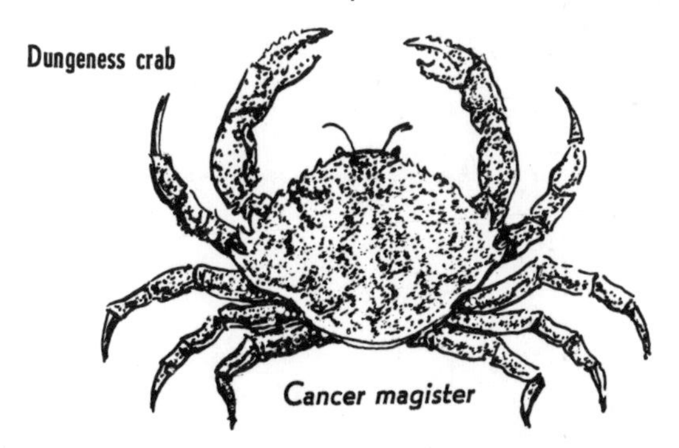

Cancer magister

The Dungeness crab is abundant along the Pacific coast of the United States and Canada, and to a lesser extent in Alaskan waters. Those found off Canada are generally in shallow water while the California animals seem to prefer depths of 6 feet or more. Many gourmets favor the Dungeness to almost any other crab on the market.

PLANKTON

All of the animals discussed so far share more than just the marine environment in which they exist. Early in their lives, whether as eggs or newly formed creatures, most of them were part of what we identify as plankton, teeming communities of microscopic plants and animals. These communities are more abundant in colder locales, but they are found in all ocean and fresh-water areas.

The list of animals which make up oceanic plankton is a long one because it includes representatives of a wide variety of alga (phytoplankton), and a multitude of vertebrate-invertebrate species (zooplankton), all of which are free floating and subject to the whims of temperature changes, upwelling, currents, tides, and winds. Included in the invertebrate zooplankton listings are the larvae of echinoderms, the eggs and young of mollusks and crustaceans, the planulae (embryos) of coelenterates, as well as protozoans, sponges and annelids.

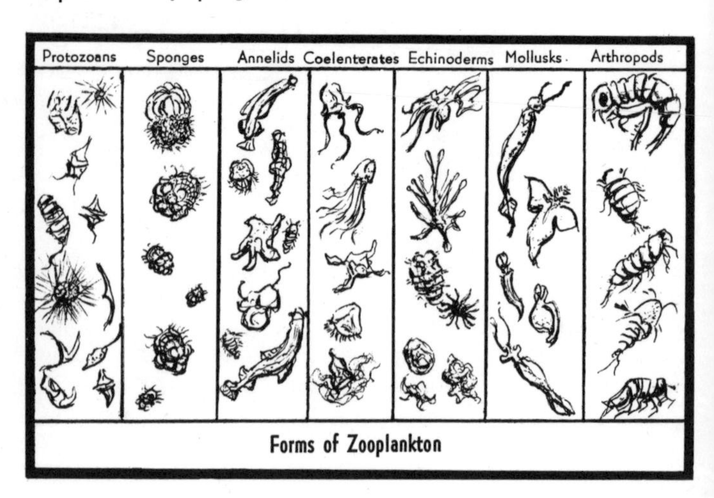

Forms of Zooplankton

Pacific and Atlantic whalers years ago recognized the importance of plankton (which they called krill) to those species of whales which feed almost exclusively on it, and marine scientists, particularly those involved in benthic studies, have long acknowledged its prime importance in the food chain for all other forms of life. Without plankton there might not be life in the ocean nor, even, on land.

Today science is divided over the possibilities of harvesting plankton for the direct feeding of human beings. Many individuals believe it to be the best immediate solution to the disturbing spectre of world hunger, while others view the idea somewhat skeptically, considering the formidable task of assembling adequate fleets of factory ships and harvesting boats (which must operate far from the continental shelf) and distribution centers on land. The question of whether humans will accept plankton remains unanswered. Also of concern are the environmental factors involved in such a massive undertaking. It must be acknowledged here, although reluctantly, that the most efficient harvester of plankton, the whale, is also the most efficient means for man to take advantage of this potentially rich source of food; each plankton-fed whale represents a multi-ton package of readily processed plankton. However, it is recognized that the gentle giants are even now facing extinction and any world-organized move to further hunt and kill them would be most ill advised.

GLOSSARY

ALGAE. Plants without roots, stems or leaves. Algae are probably the oldest of plants, and range from microscopic in size to the seaweeds hundreds of feet in length.

AMOEBAE. One-celled animals which multiply by fission. They move from place to place by forming finger-like protrusions and proceeding in a slow manner.

AMPHIPODS. Crustaceans with bodies that are flattened from side to side and divided into sections which enable them to move with unusual agility.

ATOLLS. Coral reefs, circular in shape, which form and surround lagoons.

BENTHIC. That part of the ocean inhabited by living organisms.

BENTHOS. Organisms which inhabit the ocean bottom.

CONTINENTAL SHELF. That part of the sea floor adjacent to the continent.

DIATOMS. One-celled algae whose cells include silica. The usual green coloration of diatoms is masked by a brownish pigment.

FATHOM. A measurement of depth equal to 6 feet (1.83 m).

FRACTURE ZONE. An irregular area of the sea floor which is dotted with ridges and seamounts. Most ocean basins have such zones.

GUYOTS. Seamounts which are flat on top.

INVERTEBRATES. Animals without spines or backbones.

LAVA. Liquid rock which is released by volcanos or fissures.

MAGMA. Liquid or molten rock stored within the earth's crust and mantle. (It is called lava when it is released to the surface)

MEDUSA. The jellyfish.

PELAGIC. Generally refers to open ocean areas as contrasted to coastal regions.

PHYTOPLANKTON. Plankton which are plants. (See also PLANKTON and ZOOPLANKTON.

PLANKTON. Collective term applied to all minute plants and animals that inhabit the surface of oceans or lakes.

SEAMOUNTS. Elevations extending from the sea floor. Seamounts resemble inverted cones.

SYMBIOSIS. The mutually advantageous association or living together of organisms which are dissimilar.

TIDES. The rising and falling of the level of the sea.

TOPOGRAPHY. The study of the earth's surface and its physical features.

UPWELLING. The movement of water from the deeper areas of the ocean to the surface.

VERTEBRATES. Animals with spinal columns or back bones.

ZOOPLANKTON. Plankton which are animals.

BIBLIOGRAPHY

GENERAL

Deas, W., and Lawler, C., 1970. *Beneath Australian Seas.* A. H. and A. W. Reed, Sydney. 112 pages, illustrated.

Faulkner, D., and Smith, C., 1970. *The Hidden Sea.* Viking Press, New York. 148 pages, illustrated.

Hedgpeth, J., and Hinton, S., 1961. *Common Seashore Life of Southern California.* Naturegraph Co., Healdsburg, Calif. 65 pages, illustrated.

Johnson, E. J., and Snook, H. J., 1967. *Seashore Animals of the Pacific Coast.* Dover Publications, New York. 659 pp.

Ricketts, E. F., and Calvin, J., 1962 (Edition revised by J. W. Hedgpeth). *Between Pacific Tides.* Stanford University Press. Stanford, California. 516 pages, illustrated.

Romer, A. S., 1972. *The Procession of Life.* Doubleday and Co. Inc., New York. 384 pages, illustrated.

Utinomi, H., 1965. *Seashore Animals of Japan.* Hoikusha, Osaka, 168 pages, illustrated.

ANNELIDS

Boolootian, R. A., and Heyneman, D., 1969. *An Illustrated Laboratory Text In Zoology.* Holt, Rinehart and Winston, Inc. Pages 75–90.

Dales, R. P., 1957. *Annelids.* Hutchinson University Library, London. Illustrated.

BIBLIOGRAPHY

COELENTERATES

Darwin, C., *The Structure and Distribution of Coral Reefs.* Univ. of California Press, Berkeley. 1962. 214 pages. Reprinted from "Geological Observations on Coral Reefs, Volcanic Islands, and on South America", Smith, Elder and Company. London, 1851.

Sherman, I. W., and Sherman, V. G., 1970. *The Invertebrates: Function and Form.* Macmillan, New York. Pages 57-86.

Yonge, C. M., 1958. "Ecology and Physiology of Reef Building Corals," in *Perspectives In Marine Biology.* A. A. Buzzati-Traverso, ed. University of California Press, Berkeley, Calif. Pages 117-135.

CRUSTACEANS

Edmondson, C. H., 1946. *Reef and Shore Fauna of Hawaii.* B. P. Bishop Museum, Honolulu. Pp 219-315. Illustrated.

Tinker, S. W., 1965. *Pacific Crustacea.* Charles E. Tuttle Company, Tokyo. 134 pages, illustrated.

ECHINODERMS

Furlong, M., and Pill, V., 1970. *Starfish,* methods of preserving and guides to identification. Ellison Industries, Edmonds, Washington. 104 pages, illustrated.

Nichols, D., 1962. *Echinoderms.* Hutchinson University Library, London. 200 pages, illustrated.

BIBLIOGRAPHY

MOLLUSKS

Abbott, R. T., 1968. *American Seashells.* Van Nostrand Company, Inc. Princeton, N. J. 541 pages, illustrated.

Habe, T., and Ito, K., 1965. *Shells of the World,* Vol I, the Northern Pacific. Hoikusha, Osaka. 176 pages, illustrated.

Habe, T., and Kosuge, S., 1966. *Shells of the World,* Vol II, the Tropical Pacific. Hoikusha, Osaka. 193 pages, illustrated.

Morton, J. E., 1964. *Molluscs.* Hutchinson University Library. London. 232 pages, illustrated.

Powell, A.W. B., 1957. *Shells of New Zealand.* Whitcombe and Tombs, Ltd. Christchurch, N. Z. 202 pages, illustrated.

Rippingale, O. H., and McMichael, D. F., 1961. *Queensland and Great Barrier Reef Shells.* The Jacaranda Press, Brisbane 210 pages, illustrated.

Tinker, S. W., 1952. *Pacific Sea Shells.* Charles E. Tuttle Co., Tokyo. 240 pages, illustrated.

OCEANOGRAPHY

Ross, D. A., 1970. *Introduction to Oceanography.* Appleton-Century-Crofts, New York. 384 pages, illustrated.

Turekian, K. K., 1968. *Oceans.* Prentice-Hall, Inc. Englewood Cliffs, N. J. 120 pages, illustrated.

BIBLIOGRAPHY

PLANKTON

Davis, C. C., 1955. *The Marine and Fresh-Water Plankton.* Michigan University Press. 562 pages, illustrated.

Wimpenny, R. S. 1966. *The Plankton of theSea.* S Faber and Faber, Ltd., London. 426 pages, illustrated.

PROTOZOANS

Jahn, T. L., and Jahn, F. F., 1949. *The Protozoa.* W. C. Brown Co. Iowa. 234 pages, illustrated.

Kudo, D., 1954. *Protozoology.* Charles C. Thomas, Springfield, Ill. 966 pages, illustrated.

INDEX